WILDLIFE PROTECTION, DESTRUCTION AND EXTINCTION

THE STATION FIRE

LESSONS FOR FUTURE WILDLAND FIRE MANAGEMENT

WILDLIFE PROTECTION, DESTRUCTION AND EXTINCTION

Additional books in this series can be found on Nova's website under the Series tab.

Additional E-books in this series can be found on Nova's website under the E-books tab.

ENVIRONMENTAL REMEDIATION TECHNOLOGIES, REGULATIONS AND SAFETY

Additional books in this series can be found on Nova's website under the Series tab.

Additional E-books in this series can be found on Nova's website under the E-books tab.

WILDLIFE PROTECTION, DESTRUCTION AND EXTINCTION

THE STATION FIRE

LESSONS FOR FUTURE WILDLAND FIRE MANAGEMENT

EDGAR L. DONAHUE
AND
DAWN M. ELMORE
EDITORS

Nova Science Publishers
New York

For permission to use material from this book please contact us:
Telephone 631-231-7269; Fax 631-231-8175
Web Site: http://www.novapublishers.com

NOTICE TO THE READER

Library of Congress Cataloging-in-Publication Data

ISBN 978-1-62081-084-2

Published by Nova Science Publishers, Inc. † New York

CONTENTS

PREFACE

In 2009, a wildland fire in the Angeles National Forest in California known as the Station Fire led to the death of two firefighters, destroyed 89 homes and dozens of other structures, and burned more than 160,000 acres. After escaping initial containment efforts, the Station Fire underwent periods of rapid growth and extreme fire behavior over the following several days, ultimately threatening thousands of homes in nearby communities. In response, the Forest Service and local agencies, deployed thousands of firefighters and hundreds of firefighting assets, including fire engines, helicopters, and air tankers. This book examines the key issues raised by firefighters, area residents, and others regarding the Forest Service's response to the Station Fire over the adequacy of firefighting assets, strategies, and tactics used.

Chapter 1- In 2009, a wildland fire in the Angeles National Forest in California known as the Station Fire led to the death of two firefighters, destroyed 89 homes and dozens of other structures, and burned more than 160,000 acres. The Department of Agriculture's Forest Service played a key role in managing the firefighting response. Some firefighters, area residents, and others have raised questions about how the Forest Service responded to the fire. GAO was asked to evaluate the response to the Station Fire. Accordingly, this report (1) describes key events in the Station Fire and the Forest Service's response, including strategies, tactics, and assets used; (2) examines key issues arising from this response; and (3) identifies lessons the Station Fire offers for wildland fire management in the future, including lessons specific to Southern California. GAO reviewed agency documents and interviewed officials from the Forest Service and from nonfederal firefighting agencies involved in the response, as well as other parties.

Chapter 2- On August 26, 2009, at 3:24 p.m., the Angeles National Forest (USDA Forest Service, Pacific Southwest Region) received a call for initial attack response to a fire that had ignited on the forest near the Angeles Crest Fire Station, on the Angeles Crest Highway. The terrain below the Angeles Crest Highway is rugged and steep. The slope percentage from the highway to the creek bottom ranges from 33 percent to 67 percent along the half mile section of the highway centered on the fire station. There are several narrow ravines facing east along this section of highway. The slopes below the highway range from 900 to 1,680 feet in length. The vegetation was mature mixed chaparral, standing six to eight feet tall, at least 50 years old, and extremely flammable. Ultimately, the fire grew in size to over 160,000 acres, leading to the loss of forest resources, private property, and human life. At the direction of the Chief of the USDA Forest Service, a review was undertaken for the period from initiation of initial attack on August 26 to August 28, 2009, at 2:00 p.m., when the fire was placed under the command of a Type I Incident Management Team.

Chapter 3- The Forest Service has received criticism for not utilizing night flying capability by partnering agencies during the evening of August 26 and early morning hours of the 27th , 2009. As a result, the Forest Service identified this topic as an area to explore for potential use in the future.

In: The Station Fire ISBN: 978-1-62081-084-2
Editors: E. L. Donahue and D. M. Elmore © 2012 Nova Science Publishers, Inc.

Chapter 1

STATION FIRE: FOREST SERVICE'S RESPONSE OFFERS POTENTIAL LESSONS FOR FUTURE WILDLAND FIRE MANAGEMENT*

United States Government Accountability Office

WHY GAO DID THIS STUDY

In 2009, a wildland fire in the Angeles National Forest in California known as the Station Fire led to the death of two firefighters, destroyed 89 homes and dozens of other structures, and burned more than 160,000 acres. The Department of Agriculture's Forest Service played a key role in managing the firefighting response. Some firefighters, area residents, and others have raised questions about how the Forest Service responded to the fire. GAO was asked to evaluate the response to the Station Fire. Accordingly, this report (1) describes key events in the Station Fire and the Forest Service's response, including strategies, tactics, and assets used; (2) examines key issues arising from this response; and (3) identifies lessons the Station Fire offers for wildland fire management in the future, including lessons specific to Southern

* This is an edited, reformatted and augmented version of the Highlights of GAO-12-155, a report to congressional requesters, dated December 2011.

California. GAO reviewed agency documents and interviewed officials from the Forest Service and from nonfederal firefighting agencies involved in the response, as well as other parties.

WHAT GAO RECOMMENDS

GAO recommends that the Forest Service (1) clarify when it expects agency firefighting assets be used instead of assets from other sources and (2) document the steps it plans to take in response to the lessons identified in its review of the Station Fire. The Forest Service generally agreed with GAO's findings and recommendations.

WHAT GAO FOUND

The Station Fire started on the afternoon of August 26, 2009, in steep terrain covered with highly flammable vegetation during very dry conditions. After escaping initial containment efforts, the Station Fire underwent periods of rapid growth and extreme fire behavior over the following several days, ultimately threatening thousands of homes in nearby communities. In response, the Forest Service and local agencies, such as Los Angeles County, deployed thousands of firefighters and hundreds of firefighting assets, including fire engines, helicopters, and air tankers. The fire's extreme behavior, however, often prevented firefighters from attacking it directly, instead leading them to employ tactics farther away from the fire in their efforts to protect life, homes, and natural resources. The fire was contained on October 16, 2009.

Among the key issues raised by firefighters, area residents, and others regarding the Forest Service's response to the Station Fire were questions over the adequacy of firefighting assets, strategies, and tactics used. For example, some observers questioned (1) why the Forest Service did not use certain aircraft that were available, including night-flying helicopters operated by Los Angeles County; (2) whether the agency followed appropriate procedures in ordering firefighting assets, including whether it mobilized its own assets rather than local ones in certain instances, even though its assets were located farther away and would take longer to arrive; and (3) whether more action could have been taken to protect homes in Big Tujunga Canyon, an area where dozens of homes were destroyed. GAO's review of available information was able to clarify some of these issues by, for example, identifying the location and availability of certain aircraft and other assets. In other cases, insufficient information was available to fully ascertain the facts—such as the exact procedures followed when ordering certain assets. Also, for those concerns that centered on a difference of opinion, such as whether additional actions could have been safely taken to protect homes, GAO was able to review the various perspectives of observers but had no method for addressing these differences through analysis. While some observers were critical of the Forest Service's response to the Station Fire, others commended its response, highlighting the difficult conditions confronting firefighters and the thousands of threatened homes that ultimately were protected.

The Station Fire offers several important lessons that may help improve wildland fire response in the future, including in Southern California. These lessons include the importance of (1) determining the appropriate role of night-flying aircraft, (2) having transparent processes for ordering and mobilizing firefighting assets, (3) tracking aircraft water and retardant deliveries, (4) predicting fire behavior under a variety of conditions, and (5) having systematic methods to identify needed firefighting assets. The agency has taken action to implement some lessons, including changing its night-flying policy and the asset-ordering practices at the Angeles National Forest. The agency has not, however, clarified its expectations about when its own assets are to be ordered instead of other agencies'. In addition, while the agency prepared a "lessons-learned" report after the fire, this report does not fully describe other actions the agency expects to take to implement lessons from the fire or time frames for doing so—potentially representing a lost opportunity to capitalize on the Station Fire's lessons.

ABBREVIATIONS

CAL FIRE	California Department of Forestry and Fire Protection
FPA	fire program analysis
ROSS	resource ordering and status system

December 16, 2011

The Honorable Barbara Boxer
The Honorable Dianne Feinstein

United States Senate
The Honorable Judy Chu
The Honorable David Dreier
The Honorable Howard P. "Buck" McKeon
The Honorable Adam Schiff
The Honorable Brad Sherman
House of Representatives

A wildland fire known as the Station Fire began in the Angeles National Forest in California on August 26, 2009, and, before it was declared contained on October 16 of that year, led to the death of two firefighters, destroyed 89 homes and dozens of other structures, and burned more than 160,000 acres of land in Los Angeles County. The fire, which was set by an arsonist, began during very dry conditions in steep terrain covered with highly flammable vegetation, a mix contributing to periods of rapid fire spread and difficult and dangerous conditions for firefighters. At the fire's height, more than 5,200 firefighters and other personnel responded as the fire threatened thousands of homes; critical communications and other infrastructure; and natural and cultural resources, including watersheds susceptible to postfire flooding and erosion. Suppressing the fire cost an estimated $95 million, placing it among the most costly fires in the nation's history. Because the fire began in the Angeles National Forest, the Department of Agriculture's Forest Service played a key role in managing the firefighting response. Because the Station Fire also burned and threatened cities and county land adjacent to the national forest, several local fire departments (including those from Los Angeles County and the cities of Glendale, Los Angeles, and Pasadena) entered into "unified command" with the Forest Service, meaning that those departments were involved in making decisions about how to respond to the fire.

Some firefighters, area residents, and others have raised questions about whether the Forest Service could have mobilized firefighting assets more quickly, which might have allowed the agency to contain the Station Fire sooner. Questions have also been raised about whether different firefighting strategies or tactics could have allowed the Forest Service to either contain the fire before it became large or reduce the damage it aused. In this context, you asked us to review the response to the Station Fire. Accordingly, this report (1) describes key events in the Station Fire and the Forest Service's response, including strategies, tactics, and assets used; (2) examines key issues arising from this response; and (3) identifies lessons the Station Fire offers for wildland fire management in the future, including lessons specific to Southern California.

To determine the key events of the Station Fire and the Forest Service's response, we reviewed numerous interagency incident documents, including daily incident summaries, daily plans of firefighting actions to be taken, orders for firefighting assets, and recordings and transcripts of radio transmissions. We also reviewed agency and other reports that examined particular aspects of the fire or the response to it. In addition, we reviewed Forest Service policies and guidance related to firefighting, including the 2009 *Interagency Standards for Fire and Fire Aviation Operations*, interagency guidance on mobilizing firefighting assets, and the fire management plan for the Angeles National Forest. To obtain additional information, and to provide context for the documents reviewed, we interviewed selected officials from the Forest Service and other agencies who were directly involved in the response. We also interviewed Forest Service officials from national, regional, and forest levels; officials from the nonfederal agencies involved in the response, including senior officials from the firefighting agencies that were part of the incident's unified command; and residents and others who witnessed or were affected by the fire. Some nonfederal agency officials and retired federal officials declined our requests for interviews. These officials might have provided us additional perspectives or information on the fire, but we believe that, given the number and variety of individuals we interviewed and the volume of documents we examined, our inability to interview these officials is unlikely to have substantially affected our findings, conclusions, or recommendations. In describing the Station Fire, we divided the fire into five phases to reflect changes in fire size and behavior, the risk to structures and resources, and the nature of the agency's response. To determine the key issues raised by the Forest Service's response and the lessons the Station Fire offers for future wildland fire management, we reviewed agency and other reports evaluating aspects of the response and reviewed information provided at congressional meetings held in Southern California in October 2010 and April 2011. We also discussed key issues and potential lessons during our interviews with incident personnel, Forest Service officials, officials from nonfederal agencies, residents affected by the fire, and other interested parties. Appendix I describes our scope and methodology in more detail.

It is important to note that our review had the benefit of hindsight, a benefit not available to those who responded to the Station Fire. We recognize that wildland fires can be unpredictable and that fire and weather conditions can change quickly. Firefighters have to respond to such changes as they occur, using the information they have available at a particular moment to make decisions about the strategies and tactics they believe will provide the best opportunities for success, while also considering the risk to firefighters from taking a particular action. These decisions may be made with imperfect information and under severe time constraints, relying heavily on the professional judgment of those involved. In addition, it is not possible to know with certainty whether different decisions or actions would have resulted in a different outcome for the Station Fire. In conducting our review, we evaluated decisions made and actions taken in light of the information available to firefighters and forest officials at the time.

We conducted this performance audit from December 2010 through December 2011 in accordance with generally accepted government auditing standards. Those standards require that we plan and perform the audit to obtain sufficient, appropriate evidence to provide a reasonable basis for our findings and conclusions based on our audit objectives. We believe that the evidence obtained provides a reasonable basis for our findings and conclusions based on our audit objectives.

BACKGROUND

The Angeles National Forest, along with the other national forests in Southern California, is located in one of the driest, most fire-prone areas in the United States and also abuts the major population centers of greater Los Angeles. Firefighters have suppressed fires occurring in the area for decades, which has resulted in the significant accumulation of brush and other flammable fuels in many places. Chaparral shrublands,[1] one of the most fire-hazardous landscapes in North America, dominate low to middle elevations of the Southern California foothills and mountains. Fires in this landscape tend to exhibit extreme fire behavior because the vegetation's characteristics and the steep slopes facilitate fires' rapid upslope spread. Dry and hot weather characteristic of the region also contributes to the severity of local fires. In addition, continuing development in the wildland-urban interface—where human development meets or intermixes with undeveloped wildland—has placed an increasing number of homes at risk of damage from wildland fire in this area. California has experienced a number of fires that have destroyed hundreds, and in some cases even thousands, of homes and other structures (see table 1).

The cost of suppressing wildland fires has escalated in recent years. Since 2000, numerous fires have each cost more than $75 million to suppress—including some whose suppression costs exceeded $100 million (see table 2).

Table 1. Fires in California That Destroyed the Most Structures

Name	Year	Structures destroyed (approximate)	Acres burned (approximate)
Tunnel Fire	1991	2,900	1,600
Cedar Fire	2003	2,820	273,246
Witch Fire	2007	1,650	197,990
Old Fire	2003	1,003	91,281
Jones Fire	1999	954	26,200
Paint Fire	1990	641	4,900
Fountain Fire	1992	636	63,960
Sayre Fire	2008	604	11,262
City of Berkeley Fire	1923	584	130
Harris Fire	2007	548	90,440

Source: California Department of Forestry and Fire Protection.
Note: Structures destroyed includes homes, commercial structures, and outbuildings.

Because one firefighting entity alone cannot handle all wildland fires that may burn in its jurisdiction, especially when faced with large fires like the Station Fire, agencies in the United States use an interagency incident management system that depends on the close cooperation and coordination of federal, state, tribal, and local fire protection agencies. Although the Forest Service is the predominant federal firefighting agency in terms of the amount of resources devoted to firefighting, federal and nonfederal firefighting entities generally share their firefighting personnel, equipment, and supplies and work together to fight fires, regardless of which entity has jurisdiction over the burning lands. Agreements between cooperating entities govern these cooperative fire protection efforts and contain general provisions for sharing firefighting assets and costs.

On a large wildland fire, fire suppression efforts generally fall into two phases—initial attack and extended attack. The initial attack phase consists of the efforts to control a fire during the first "operational period" after the fire is reported, generally 24 hours.[2] Local fire management officials—for example, a fire management officer for a national forest— direct these initial firefighting efforts in their jurisdictions. While the majority of fires on Forest Service land—about 98 percent—are controlled and suppressed during initial attack, a small percentage escape and require further firefighting efforts. Such additional efforts are referred to as extended attack.

Table 2. Fires in the United States with the Highest Suppression Costs, 2000-2011

Name	Year	State	Suppression cost (approximate, in millions)	Acres burned (approximate)
Biscuit Fire	2002	Oregon, California	$152.7	499,570
Klamath Theater (Siskiyou Complex, Bear Wallow Complex, Panther Fire)	2008	California	126.1	192,038
Zaca Two Fire	2007	California	122.5	240,207
Wallow Fire	2011	Arizona, New Mexico	109.0	538,049
Station Fire	2009	California	95.5	160,577
BTU Lightning Complex	2008	California	94.8	64,995
Tripod Complex	2006	Washington	84.1	175,184
Basin Complex	2008	California	78.1	162,818
Day Fire	2006	California	78.0	162,702
Texas Winter Fires	2008	Texas	77.7	22,137

Source: Forest Service.

Notes: Dollars are given as reported by the Forest Service and are not adjusted for inflation. According to a Forest Service official, data from fires before 2000 are not readily available, but it is unlikely that suppression costs for earlier fires exceeded those of fires occurring since 2000.

During initial attack, firefighting entities normally follow a principle of closest available resource, whereby, regardless of jurisdiction, the closest available firefighting personnel and equipment respond. As the fire moves into the extended attack phase, officials consider a number of factors— including whether the asset will arrive at the time requested—and order the assets they believe are most appropriate, which may not necessarily be the closest. Personnel, aircraft, equipment, and supplies are ordered through a system of local, regional, and national dispatch centers. If assets are insufficient in the local dispatch area close to a wildland fire, dispatch center personnel forward requests to the responsible regional dispatch center, which locates and sends additional firefighting assets from elsewhere within the region.[3] If necessary, the regional dispatch center can forward requests to the National Interagency Coordination Center in Boise, Idaho, which locates and assigns the closest available asset to the fire.

The Forest Service and its interagency firefighting partners use an incident management system designed to provide appropriate management and leadership team capabilities for firefighting efforts. The fire's complexity determines the type of leadership team and firefighting assets assigned. There are five types of incidents, increasing in complexity from type 5 (the least complex) to type 1 (the most complex). For example, to manage a type 5 incident, the incident commander may be a local employee qualified to direct initial attack efforts on a small fire with two to six firefighters. In contrast, for a type 1 incident, the incident commander is one member of a highly qualified management team of firefighting personnel that often number more than 500. Incident management teams are also rated on a scale according to their training and experience, with type 1 teams qualified to handle the most-complex incidents. Type 1 team members receive additional training and generally have the most experience in handling complex incidents. A single incident commander is typically in charge, but the incident management system may be expanded into a unified command structure when multiple jurisdictions are involved. This structure brings together commanders of the relevant jurisdictions as a way to facilitate a coordinated and integrated response. In such cases, members of the unified command work together to develop a common set of incident objectives and strategies, maximize the use of assets, and enhance the individual response organizations' efficiency.

Ultimately, the team that is assigned works with Forest Service line officers—such as forest supervisors or district rangers—and fire management staff from the affected national forest to determine the strategies and tactics to use in managing a fire. The strategy is the overall plan designed to control the fire; for example, a strategy may be to protect threatened structures and to contain the fire within a certain geographic area. Tactics are actions that are taken to accomplish the objectives set out in the strategy. For example, the fire may be attacked directly, with firefighters working at the fire's edge to extinguish it. If the fire is burning too intensely for direct attack or if direct attack is otherwise not possible, practical, or safe—because the fire is burning on very steep slopes, for example—firefighters may choose to attack it indirectly. In such cases, firefighters typically select an area away from the fire and construct a "fireline," where vegetation is cleared in an effort to stop the fire's spread at that point or slow it sufficiently to allow firefighters to attack directly. Firefighters often incorporate geographic features such as roads, rocky areas, or ridgelines into firelines to increase their effectiveness. In some cases firefighters will conduct backfiring operations, in which they intentionally set fire to fuels between a fireline and the fire to slow or contain a rapidly spreading fire by depriving it of fuel.

In carrying out these strategies and tactics, firefighters use a variety of firefighting assets, both on the ground and in the air. Ground-based assets include firefighting crews; wildland fire engines; and machinery such as bulldozers, which firefighters use to help them construct firelines. Aviation assets include helicopters, which can drop water directly on a fire, and fixed-wing air tankers, which can drop fire retardant ahead of the fire, often near a fireline that has been constructed, to slow a fire's spread. (Fire retardant is most effective when applied ahead of the fire and is not applied directly on a fire.) Less commonly used assets include certain fixed-wing aircraft that carry water or suppression gel that is dropped directly on a fire. Fixed-wing air tankers range in size from small single-engine planes, which are maneuverable but carry only small amounts of retardant, to a limited number of large aircraft such as converted DC-10s or Boeing 747s—referred to as "very large air tankers"—which can carry substantial amounts of retardant but whose use can be limited in mountainous terrain because of their size.[4]

Firefighters and other personnel who respond to wildland fire incidents are required to complete training to help them identify hazards, as well as appropriate strategies and tactics to respond to different situations. The level of risk that decision makers and firefighters are willing to accept in any given situation—and, as a result, in some cases the strategies and tactics ultimately used—depends on the experience of the firefighters involved. Overall, agency firefighting doctrine emphasizes safety above all other concerns; Forest Service policy, for example, states, "In conducting wildland fire suppression, responsible officials shall give first priority to the safety of firefighters, other personnel, and the public." In determining which firefighting strategies to follow, fire managers are instructed to make safety their primary criterion.

Key Events of the Station Fire and the Forest Service's Response

The Station Fire started on the afternoon of August 26, 2009. After escaping initial containment efforts, the fire underwent periods of rapid growth and extreme fire behavior during the following several days. In response, the Forest Service and other agencies deployed thousands of firefighters and hundreds of fire engines, helicopters, air tankers, and other assets. The fire's extreme behavior, however, often prevented firefighters from attacking it directly, instead leading them to use indirect tactics to protect life, homes and communities, and natural resources. Ultimately, the fire burned more than 160,000 acres and threatened thousands of homes and other structures (see figure. 1). The following sections describe key events of the Station Fire and the Forest Service's response during each of five phases into which we have divided the fire. Appendix II contains maps detailing the areas that burned on particular dates.

Phase 1 (August 26)

The Station Fire was set by an arsonist on the afternoon of August 26, near a pullout on the Angeles Crest Highway.[5] According to a Forest Service report, the fire began in approximately 50-year-old chaparral about 6 to 8 feet high and containing more than 50 percent dead material—conditions that, when combined with steep slopes, low humidity, and high temperature, were conducive to long flame lengths,[6] intense burning, and rapid spread. A "red flag" warning was in effect for the area that afternoon, indicating weather that could result in extreme burning conditions. Fire records indicate that within about an hour of being reported, the fire had burned the hillside above the highway pullout and covered approximately 15 to 20 acres.

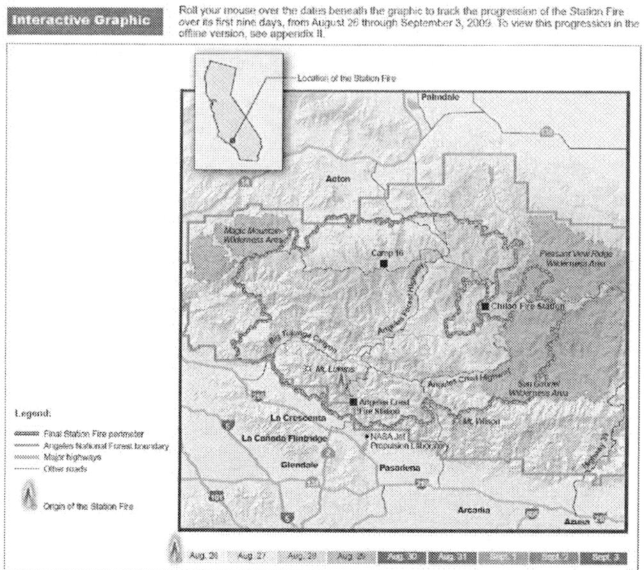

Sources: GAO analysis of Forest Service information; Forest Service (map).

Figure 1. Angeles National Forest and Vicinity, with Perimeter of the Station Fire and Daily Fire Progression.

Source: GAO (photo) and Forest Service (fire information).

Figure 2. Location of the Station Fire on August 26, 2009, Showing Firefighter Divisions Alpha and Zulu.

The fire was reported at about 3:15 p.m., and firefighters from both the Angeles National Forest and Los Angeles County responded. The first firefighters arrived at approximately 3:30 p.m., and by 4:30 p.m. approximately 175 firefighters and other personnel and 14 engines were on the scene, according to agency officials. In addition, seven helicopters and two air tankers were mobilized to respond to the fire during the afternoon, with several aircraft arriving over the fire before 4:00 p.m. These resources were under the overall command of an incident commander assigned by the Angeles National Forest. According to forest officials, the initial firefighting response to the Station Fire was greater than the typical response to a new wildland fire, because aircraft were able to arrive on scene more quickly than usual from another fire, the Morris Fire, which had started nearby in the Angeles National Forest the day before. As a result, the Forest Service was able to quickly move assets from that fire to the Station Fire.

In accordance with the Angeles National Forest Fire Management Plan, the overall objective in responding to the Station Fire during this phase— as with all fires burning in the Angeles National Forest—was full suppression.[7] To accomplish this objective, firefighters attempted to create firelines around the fire in an effort to contain it in the immediate area above the Angeles Crest Highway.[8] The attack on the fire was divided into two divisions: Los Angeles County firefighters were assigned to Division Alpha, on the fire's left flank, while Forest Service firefighters were assigned to Division Zulu, on the right flank (see figure 2). Firefighters told us that Division Alpha's county firefighters first attempted a direct attack on the fire's left flank, but the fire was too intense, and the division supervisor determined that the firefighters should fall back and try instead to establish an indirect fireline in the next drainage, approximately 200 yards to the west. Meanwhile, Division Zulu's Forest Service firefighters began to create a fireline and place fire hoses along a ridgeline close to the fire. A firefighter present at this time estimated the flames were mostly about 8 to 10 feet long; agency firefighting doctrine indicates that flame lengths greater than 4 feet are not safe for firefighters using hand tools to attack directly.

The ground-based firefighters were supported by seven helicopters and two air tankers during the afternoon and evening of August 26. According to Forest Service officials and fire records we reviewed, helicopters dropped water either on or very near the fire, which helped reduce the fire's intensity and allowed firefighters to work to establish firelines closer to the fire. In addition, air tankers dropped fire retardant on the slopes and ridgelines above the fire, in an effort to keep the fire from spreading beyond these areas.

Aircraft made numerous drops throughout that afternoon and evening. The two air tankers, for example, dropped about 15,000 gallons of retardant, according to fire records, and Forest Service officials estimated the seven helicopters dropped more than 142,000 gallons of water. Because aircraft are generally not allowed to fly firefighting missions at night,[9] the aircraft began to make their last drops on the fire around 7:00 p.m., according to the incident commander, and the last helicopters were released at 7:55 p.m.[10]

The incident commander and other firefighters who responded to the fire on August 26 told us that the steep slopes and active fire behavior made it difficult to establish firelines and contain the fire but that firefighters made progress throughout the afternoon and evening and that, although the fire was not contained by early evening, the firelines surrounding it appeared to be holding. At this time, firefighters had made sufficient progress that the incident commander believed they would be able to contain the fire that night. According to the incident commander and other forest officials, the number of firefighting assets he ordered for the next day was commensurate with "mop-up" operations—meaning that the fire was not expected to be spreading and that firefighters would be focusing on reinforcing the firelines and cooling off hot spots within the perimeter. Between approximately 10:30 p.m. and midnight, the incident commander released a number of Forest Service and county firefighting assets he believed were no longer needed to fight the fire. At about the same time, command of the fire transferred from the type 3 incident commander who had been managing the response to a type 4 incident commander.

The indirect attack on the left flank of the fire, however, left about one-quarter acre of unburned vegetation remaining, generally referred to as the "green island." Around 11:30 p.m., as down-canyon winds began to blow embers from the green island across the highway, small "spot" fires ignited in the steep canyon below the highway. Several firefighters told us that it is common for down-canyon winds to develop after sunset, and given that the green island was still burning, they had expected that embers would start to land below the highway at some point during the night. Over the next 2 to 3 hours, embers ignited at least six spot fires, according to fire records.[11] Firefighters involved in the response told us they were able to suppress several of these spot fires, but one of them was burning in a location that was too dangerous to attack. Firefighters said these spot fires posed a substantial risk to firefighters because they were on steep slopes. It is often unsafe to attack a fire on steep slopes from above because fire can burn uphill fast; it can also be unsafe to attack a fire from below because burning material can roll downhill past firefighters and ignite vegetation, which can then burn back uphill toward them. Moreover, firefighters told us the risk from these conditions is greater at night and in dense vegetation because it is difficult to see the terrain and the fire.

Because of the fire's increased severity and the higher risk posed to firefighters by the spot fires below the highway, command of the fire transferred from the type 4 incident commander back to the type 3 commander at 2:29 a.m. on August 27, although both officials told us the type 3 incident commander was present and had offered advice to the type 4 commander beginning around midnight. Once firefighters determined they were unable to fully suppress the spot fires below the road, the type 3 incident commander ordered additional aircraft to help suppression efforts in the morning. Specifically, at about 12:30 a.m. on August 27, the incident commander radioed an Angeles National Forest dispatcher and requested three air tankers to be over the fire at 7:00 a.m. He also ordered a third helicopter, in addition to two he already had ordered earlier in the evening.

Phase 2 (August 27 through Afternoon of August 28)

On the morning of August 27, as firefighters renewed their efforts, the overall goal was to contain the Station Fire before it could reach nearby communities and infrastructure, according to one of the firefighters supervising the response. Incident officials and firefighters told us they recognized the importance of attacking the fire as early as possible during the day because it was burning on east-facing slopes and would become more active through the day as the sun rose and warmed the land. Four air tankers responded to the Station Fire during the morning, arriving at the fire at approximately 9:00 a.m., and seven helicopters also responded, arriving at the fire between approximately 7:45 a.m. and 10:30 a.m.[12] A firefighter supervising the response in that area told us that, in an effort to keep the fire from spreading, he had air tankers drop retardant on nearby ridges (which were not burning) and had helicopters drop water directly on the fire in an effort to slow its progress and reduce its intensity to allow firefighters more time to try to create firelines around it. In addition, ground crews continued to attack the fire, including a crew of "hotshots"—18 to 20 highly qualified firefighters—working to create a fireline to keep the fire from spreading. As the fire intensified during the day, however, crews had to abandon this effort. The firefighter supervising the response emphasized the fire's intensity, noting that as early as 7:45 a.m., water dropped by helicopter was having little effect;

this firefighter also told us that the fire burned through retardant that had been dropped by air tankers that morning.

These efforts notwithstanding, the fire burning below the highway—which had started from the embers from the green island during the night—began to spread south on the morning of August 27, paralleling the Angeles Crest Highway toward the mouth of the canyon and densely populated areas adjacent to the forest. The fire also began to move up the canyon slopes toward the highway, ultimately crossing it once again; now above the highway, the fire subsequently spread quickly both up and down the canyon to the north and south. (See figure 1 and app. II for maps of the fire's progress.) At this point, the fire posed a direct threat to nearby developed areas, including homes in La Cañada Flintridge, the National Aeronautics and Space Administration's Jet Propulsion Laboratory, and major electrical infrastructure. Incident documents show that by the evening of August 27, approximately 500 homes were threatened by the fire, a number that increased to approximately 1,800 by the afternoon of August 28. According to firefighters, protecting these areas was the priority for the evening of August 27 and on August 28. They told us they had air tankers drop retardant near the homes adjacent to the forest boundary and stationed engines near these homes to respond if needed.[13] They also told us they had firefighters and engines try to attack the fire directly at various times but that the fire's intensity, combined with the steep terrain, made it unsafe for firefighters to engage too close to the fire. On August 28, they said that engines and hotshot crews were also used to try to keep the fire from spreading farther east or north, to protect communications infrastructure on nearby Mt. Lukens, as well as residences in Big Tujunga Canyon to the northwest. Although firefighters were unable to contain the fire, no homes were destroyed during this phase.

As the complexity of the incident escalated, the management structure of the response changed, and the number of assets assigned to the fire rose. On the morning of August 27, the incident was managed by a type 3 incident commander. During that morning, Angeles National Forest officials, in conjunction with firefighters assigned to the incident, began discussing the need to order a type 1 incident management team; forest officials ordered this team shortly before 2:00 p.m. on August 27. But because the type 1 team was not expected to assume command until the evening of August 28, officials decided to make the Station Fire a branch of the nearby Morris Fire—meaning that the type 2 incident management team managing the Morris Fire would be responsible for managing the Station Fire as well. Ultimately, the type 1 team assumed command of the Station Fire at 2:00 p.m. on August 28. In addition, because the fire threatened lands under the jurisdiction of both the Angeles National Forest and Los Angeles County, the Forest Service entered into unified command with the county at 1:10 p.m. on August 27, so that county officials were more formally involved with making decisions about how to manage the fire.[14] During this phase, the number of firefighting assets assigned to the fire increased from an estimated 191 firefighting personnel, 4 air tankers, 7 helicopters, and 6 engines on the morning of August 27 to an estimated 637 firefighting personnel, 8 air tankers, 7 helicopters, and 40 engines on the afternoon of August 28. By that afternoon, the fire had grown to approximately 5,000 acres.

Phase 3 (Afternoon of August 28 through August 31)

On Friday, August 28, at 2:00 p.m., a type 1 incident management team assumed command of the Station Fire.[15] This command transition occurred in the middle of an operational period, generally considered a dangerous time to change command. As the evening progressed, the fire moved quickly to the south, west, and east and continued to threaten more than 1,800 homes, particularly in the foothill communities of La Cañada Flintridge and La Crescenta. Firefighters noted that the fire was moving in multiple directions simultaneously, making it difficult for them to predict where it was headed and forcing them to take firefighting actions in multiple areas. As a result, members of the incident management team told us they chose to focus their efforts near La Cañada Flintridge and La Crescenta—the area facing the greatest risk from the fire that night. Team members told us they continued to use air tankers to drop retardant to try to slow the fire's progress toward these communities and helicopters to drop water in an attempt to lessen the fire's intensity and provide opportunities for firefighters to contain the fire; parts of La Cañada Flintridge were evacuated during the night of August 28. A red flag warning remained in effect throughout this phase; moreover, although wildland fires typically become less active at night when temperatures are lower and humidity higher, which allows firefighters to make progress, firefighters told us that the Station Fire continued to burn actively at night during this phase.

On August 29, the Station Fire continued to spread in several directions. The fire continued to threaten the foothill communities to the south, west, and east and was also moving toward Mt. Wilson—the site of critical communications infrastructure—and Big Tujunga Canyon. The overall strategy continued to be full suppression, but, according to incident management officials, extreme fire behavior provided little opportunity for firefighters to attack the fire directly. Throughout the day, incident officials said they used firefighting assets where they believed doing so would be safe and effective. For example, because the fire was still threatening communities along the foothills, and mandatory evacuations were issued for portions of Altadena, La Cañada Flintridge, La Crescenta, and Glendale, incident management officials said they continued to use helicopters and air tankers—including very large air tankers, which can drop substantially more retardant than typical air tankers[16]—as well as firefighting crews and bulldozers to try to establish firelines to protect these communities.[17] By the evening of August 29, the fire threatened an estimated 10,000 homes, and approximately 1,800 personnel had been assigned to the incident.

In addition to the populated foothill communities, the Station Fire also threatened homes located within the boundaries of the Angeles National Forest. In particular, the fire threatened Big Tujunga Canyon, which had about 200 homes, including homes on private property within the national forest and recreational residences on lands leased from the Forest Service. Big Tujunga Canyon lies on the other side of a ridge from where the fire started. The fire had entered the upper reaches of one of Big Tujunga Canyon's side canyons no later than the afternoon of August 28, according to firefighters, Angeles National Forest officials, and canyon residents, but firefighters and forest officials said the fire was not intense and was burning slowly downhill—a low-intensity fire known as a backing fire. On the evening of August 28, after taking over management of the fire, the type 1 incident management team, in conjunction with officials from the other agencies in unified command, created a "structure protection group," consisting of three "strike teams" of five engines each, to protect the homes in the canyon. [18] The leader of the structure protection group told us he toured the canyon during the evening of August 28 and, in consultation with an Angeles National Forest official familiar with the area, determined that if the fire were to reach an area known as the Grizzly Flats plantation, residents of the canyon should be evacuated.

The next morning, on August 29, he returned to the canyon and noted that the fire had not changed much overnight and was still burning in the upper reaches of the canyon and slowly backing down. He told us he discussed the fire with Angeles National Forest officials who were present in the canyon, and they all agreed the fire was likely to continue burning and would threaten homes in Big Tujunga Canyon, although likely not for another 24 to 48 hours. The general consensus of the group was that when it did burn through the canyon, the fire would be of moderate intensity. A strike team leader told us that the structure protection group leader relayed this information to the firefighters in his group and directed them to take actions in the meantime, such as clearing vegetation, to help create defensible space around the homes in the canyon.[19]

Contrary to these expectations, however, the intensity of the fire in Big Tujunga Canyon increased dramatically on the morning of August 29 and quickly became extreme. The leader of the structure protection group said that the lifting of a temperature inversion that had prevailed throughout the morning had intensified the fire by increasing the flow of oxygen to it and had also allowed the heat and smoke from the fire to rise higher in the atmosphere.[20] As the fire began burning hotter, it produced more heat and smoke, and several substantial columns of smoke developed, reaching altitudes of 20,000 to 40,000 feet. As one of these columns began developing near Big Tujunga Canyon, the fire began to spread more quickly, burning through the Grizzly Flats plantation—the trigger point for evacuating the canyon—and an evacuation order was issued at approximately 10:20 a.m., according to incident management officials.[21]

Incident management officials said that fire conditions—in particular, the substantial smoke column—led to dangerous "plume dominated" fire behavior, which is atypical behavior in Southern California, according to local firefighting officials. The column subsequently collapsed, creating high winds of very hot air blowing from the column's base and pushing the fire very quickly in many directions, including down the canyon toward homes. Firefighters who were in the canyon that day and incident management officials told us that the change in fire behavior happened very quickly and that the fire reached the homes in just a few hours, much faster than the 24 to 48 hours they had expected, leaving them with few options to respond. One of the strike team leaders said that the firefighters under his command were still removing vegetation and taking other actions to help protect homes from the fire, but as the fire quickly moved down the canyon, he had to order them to retreat to a safety zone where they could wait as the fire burned around them.

After the fire burned through the area, firefighters who were in the canyon told us their first priority was to locate any civilians who had not evacuated and had stayed in the canyon. Firefighters provided emergency medical care to three civilians who sustained serious burns until they could be evacuated. Firefighters also reported they took actions to protect homes that may have ignited during the fire but were not completely destroyed or that may have still been threatened from burning materials close by. A report by one of the strike teams assigned to the area stated that they were able to save several homes through such actions.[22] Nonetheless, the Station Fire destroyed 60 residences as it burned through Big Tujunga Canyon.

Several experienced firefighters told us that fire behavior in Big Tujunga Canyon was the most severe they had seen in their careers. The leaders of both the structure protection group and one of the strike teams told us that, given the fire's unpredicted and extremely rapid escalation into the canyon, they felt fortunate that no firefighters had been killed by it. Several experienced firefighters told us that, in their opinion, the firefighters working in the canyon quickly identified that conditions had changed and made the right choice to retreat to the safety zones when they did. In all, August 29 resulted in the burning of almost 32,000 acres, bringing the total area affected by the fire to almost 37,000 acres.

Extreme fire behavior continued on August 30, with substantially more land burning on that date: approximately 57,000 acres, bringing the total area burned to about 94,000 acres. The Station Fire continued to move in multiple directions, threatening a growing number of foothill communities, including Glendale and Pasadena to the south, and making a strong push north toward Acton and east toward Mt. Wilson. The fire also burned several residences and commercial structures located within the forest boundary, including several along the Angeles Forest Highway. Firefighters anticipated extreme fire behavior similar to that of the day before and, as additional communities became threatened, focused firefighting efforts on protecting them. Incident officials said that efforts to protect communities included using bulldozers to create firelines and air tankers to drop fire retardant, including five retardant drops from the DC- 10 very large air tanker, according to incident records. In addition, three Los Angeles County Fire Department helicopters were used during the night in the Acton area. By the end of the day, an estimated 2,575 personnel had been assigned to the fire.

As the fire burned north on August 30, it moved toward a Los Angeles County Fire Department facility known as Camp 16,[23] where 74 firefighters were located. Two Los Angeles County firefighters—Fire Captain Ted Hall and Fire Fighter Specialist Arnie Quinones—lost their lives as they fought to protect the camp, and 13 firefighters were injured. The Los Angeles County Fire Department was in unified command with the Forest Service at this time, but the firefighters in the camp were not assigned to the Station Fire, and the county did not notify the incident management team of their presence at the camp, according to both a Los Angeles County report and incident management team officials we spoke with.[24]

The next day, August 31, the fire was still active and continued to threaten thousands of homes, but incident officials told us its behavior began to moderate, burning significantly less acreage (about 7,600 acres on that day) than during the previous 2 days. Nonetheless, plume- dominated conditions continued in some areas, particularly to the north and west, and mandatory evacuations continued for several communities, including parts of Acton. Incident officials said that actions that day included using crews and bulldozers to construct fireline and conducting backfiring operations. Air tankers, including very large air tankers, also continued to drop retardant in an effort to protect homes and contain the fire. Specifically, the DC-10 made eight drops and the 747 made two drops. By the end of this phase, an estimated 3,655 personnel had been assigned to the fire, which had burned over 100,000 acres in total.

Phase 4 (September 1 through September 14)

The fire continued to burn actively during the early part of this phase, burning over 13,000 acres on September 1 and 21,000 acres on September 2—which brought the total acres burned to about 137,000 acres by the end of September 2. After that date, although the fire was still expanding and threatening structures, its severity moderated as cooler, moister weather set in, and containment efforts showed increasing success. One firefighting official characterized this period as fairly standard firefighting when compared with the chaos and rapid growth of the fire's first week. As the fire's intensity moderated, firefighters were increasingly able to attack the fire directly, using aircraft to drop retardant and crews and bulldozers to construct firelines. By September 4, firefighters had completed fireline construction on the fire's western flank and were also making progress on the fire's northeastern perimeter. Where the fire still threatened homes or was too intense for firefighters to attack directly, incident officials told us they continued to use aircraft, including very large air tankers, to protect homes, communities, and infrastructure. For example, on September 4, the 747 made three drops, some of which were made near Chilao, an area with several residences threatened by the fire. From September 2 to September 7, the DC-10 was used on all but one day. An incident management team member credited one such drop, made on September 5, with saving the Angeles Crest Christian Camp, a private camp within the forest boundary.

During this phase, the fire spread farther east, expanding its threat not only to foothill communities such as Arcadia and Sierra Madre, but also to critical communications equipment atop Mt. Wilson and the surrounding area. Efforts to protect Mt. Wilson included creating defensible space around the facilities, constructing firelines, dropping substantial amounts of water and retardant by helicopter, and delivering a 5,000-gallon container of retardant for additional structure protection, according to incident officials. A Martin Mars air tanker was also used to help protect the area on September 1.[25] By September 12, operations intended to protect the area around Mt. Wilson were completed.

More broadly, beginning about September 4, cooler weather allowed for direct attack on all divisions of the fire. By September 8, more moderate fire behavior allowed for direct attack and construction of firelines in both the Pleasant View Ridge Wilderness Area to the northeast and the San Gabriel Wilderness Area to the east. In particular, incident management team members told us that they used an aggressive strategy, including using hotshot crews to build firelines through the San Gabriel Wilderness Area to keep the fire from burning too far into the wilderness. Success here was important, they told us, because the incident management team had been warned that once the fire burned into the wilderness, there was likely little that could be done to prevent it from burning all the way across to the wilderness's eastern boundary.

Additional assets continued to be assigned during this phase, reaching a peak of approximately 5,200 personnel on September 4, before declining to about 1,190 on September 14. By the end of this phase, the Station Fire had burned nearly 160,000 acres, almost its full extent. At the end of this phase, on September 14, the fire was transferred to the control of a type 2 incident management team, reflecting the lower threat.

Phase 5 (September 15 through December 4)

Throughout this phase, firefighters focused on reinforcing the fire's perimeter to ensure the fire did not rekindle. For example, ground crews repaired and maintained firelines constructed earlier to ensure that the lines would hold. In addition, because Santa Ana winds—strong winds that are common to Southern California during autumn—were predicted to develop beginning on September 19, the incident management team ordered additional aerial assets to reinforce firelines and cool off hot spots within the fire's perimeter. Incident records show that four large helicopters were ordered specifically to extinguish remaining hot spots and reduce the risk of embers igniting additional fuels, some of which were located in steep, rugged terrain that was not easily or safely accessible to ground crews. In addition, because some hot spots were located on a north-facing slope below Mt. Wilson, and the fire had not burned sufficient ground fuels for the location to be considered secure, the team used the Martin Mars air tanker to make additional drops there. During this phase, the team also took action to mitigate the damaging effects of fire suppression activities on natural resources—by, for example, restoring areas where bulldozers had been used to create firelines—so as to reduce future problems with erosion, which could lead to long-term damage in severely burned watersheds.

As the fire came under control during this phase, the number of firefighting personnel declined from approximately 980 on September 15 to 350 on September 28. After September 28, the threat from the fire had decreased to the point where command of the incident was transferred to a type 3 incident management team from the Angeles National Forest. Forest officials declared the Station Fire contained on October 16,[26] and the fire, having burned a total of more than 160,000 acres, was declared extinguished on December 4.

THE FOREST SERVICE'S RESPONSE TO THE STATION FIRE RAISED ISSUES RELATED PRIMARILY TO THE FIREFIGHTING ASSETS, STRATEGIES, AND TACTICS USED

Among the key issues raised by firefighters, area residents, and others in relation to the Forest Service's response to the Station Fire were questions over adequacy of firefighting assets, response strategies and tactics, and overall administration of the response. We grouped the issues raised by these observers into 11 categories, which are discussed in detail below. For some of these issues, we were unable to fully answer the questions raised, either because the issue centered on a difference of opinion not resolvable by analysis or because available information was insufficient to fully answer the question. It is important to note, however, that although some observers criticized certain aspects of the Forest Service's response, many others commended its response, highlighting the difficult conditions confronting fire-fighters and the thousands of threatened homes that were ultimately protected.

Martin Mars Air Tanker Not Used on August 26

Some observers questioned why the Forest Service did not use the Martin Mars air tanker on the first day of the Station Fire, stating that the suppression gel the tanker carried could have helped put out the fire while it was still small; the tanker was already under contract to the Forest Service, and that day it was nearby and available to the agency. On the basis of our review, we determined that around 6:30 p.m. on the evening of August 26, the Martin Mars was preparing to drop suppression gel on the nearby Morris Fire. According to both the pilot of the Martin Mars and the official in charge of directing the aircraft that day,[27] visibility over the Morris Fire was poor because of heavy smoke, and the aircraft was unable to make its intended drop. The official in charge of directing the Martin Mars then contacted the aviation supervisor for the Station Fire to inquire if that supervisor would like the Martin Mars to fly to the Station Fire and drop its gel on that fire instead.

Accounts of the resulting conversation differ. According to the official directing the Martin Mars, the aviation supervisor for the Station Fire responded that firefighting efforts on the Station Fire appeared to be working and that he did not have a drop location selected, but he would find a place for the aircraft to make a drop if the Martin Mars official wanted him to do so. The official directing the Martin Mars told us he decided not to have the tanker fly to the Station Fire and make the drop there because no drop location had been selected, and it was not clear that there was sufficient time for the aircraft to fly to the Station Fire, make the drop, and return to its base before nightfall.[28] He said that without a clear threat from the fire or a location identified for the drop, he did not believe the benefits of making the drop outweighed the risk of returning to base after nightfall.

In contrast, the Martin Mars pilot told us that nightfall was not a limiting factor and that he believed he had ample time for the tanker to fly to the Station Fire, make its drop, and return to base before nightfall. He told us that, on the basis of the radio conversation he could hear between the two aviation officials, he did not believe that the Station Fire aviation supervisor offered to find a location for the Martin Mars to drop its load; instead, he told us that the Station Fire aviation supervisor declined the Martin Mars because it would interfere with the water-dropping helicopters already over the fire. We were unable to interview the Station Fire aviation supervisor, who is retired from the Forest Service, and so were unable to obtain his perspective on the decision not to use the Martin Mars, including whether he believed other safety considerations were associated with its use at the time. Ultimately, the

Martin Mars remained in the vicinity of the Morris Fire, dropping its load on a nearby ridge at 6:55 p.m. and returning to its base at 7:15 p.m.

No Nighttime Helicopter Use on August 26

Some observers questioned why the Forest Service did not use Los Angeles County's night-flying helicopters during the first night, which they believe might have allowed firefighters to contain the fire. On the basis of our review, we determined that on the afternoon of August 26, the incident commander, in consultation with a Los Angeles County Fire Department official, ordered one of the county's night-flying helicopters. The incident commander told us that he requested the helicopter because he planned to use it to support firefighting operations through the end of daylight hours, operations that required a night-flying helicopter because the aircraft would return to base more than 30 minutes after sunset. According to a county dispatch log, the county was notified at 5:36 p.m. that the Station Fire was requesting a helicopter for nighttime operations; the county dispatched the helicopter to the fire at 5:43 p.m. but then recalled it at 6:01 p.m. because it was needed for emergency medical service coverage.

Both Forest Service and Los Angeles County Fire Department officials told us that the recall of this helicopter did not affect the response to the Station Fire, noting that three other county helicopters with night-flying capability remained assigned to the fire and that if the incident commander had determined that night-flying operations were needed, he could have requested to use one of those helicopters. The incident commander and other Forest Service officials told us that, sometime between 8:30 and 10:00 p.m., they discussed this possibility with the county official present at the fire and collectively they decided the gain from flying at night was not sufficient to warrant the risk because at the time the fire did not pose an immediate threat to life or property. This decision is consistent with Forest Service practice, because according to a senior Forest Service official, the agency has generally concluded that the risks of flying at night outweigh the benefits.

Release of Firefighting Assets on August 26

Some observers questioned why the Forest Service released some fire engines and firefighting crews on the first evening of the fire, stating that the fire was not yet contained and those assets would have been beneficial later that night when the fire became more active. On the basis of our review, we determined that between 10:00 p.m. and midnight on August 26, the type 3 incident commander released five Forest Service and three Los Angeles County fire engines, and one Forest Service and five Los Angeles County hand crews.[29] The incident commander, in consultation with Los Angeles County officials at the fire, determined that these assets were not needed because the fire was not spreading at that time, and he wanted the crews to be rested and available the next day. Forest Service officials told us that even with the release of these assets, approximately 61 firefighters remained at the scene during the night. Around 11:30 p.m., the fire became more active as it burned through the green island, and embers began starting spot fires below the Angeles Crest Highway. The incident commander told us that it is not clear whether additional firefighting personnel at this time would have been helpful, given that the greatest threat came from the spot fires below the highway, and officials on the scene had already decided it was too dangerous for firefighters to attack some of these spots at night. Once it became apparent that firefighters would be unable to extinguish those fires, the incident commander ordered more firefighting assets—including six firefighting crews and one helicopter—for the next morning, in addition to the assets he had ordered earlier in the evening.

Delayed Arrival of Air Tankers on August 27

Some observers questioned why three air tankers that were ordered around 12:30 a.m. on the first night of the fire arrived about 2 hours later than requested by the incident commander—suggesting that firefighters might have been able to contain the fire had the tankers arrived at the time requested. Moreover, questions have been raised about whether the dispatcher responsible for placing the order for the tankers processed the request improperly or whether the incident commander's request was canceled or delayed by other Angeles National Forest officials. We found that the events surrounding the ordering of the air tankers were not fully documented, and it is therefore not possible to determine the exact sequence of events in the ordering process. Nonetheless, on the basis of the information we reviewed, it appears unlikely that the requested aircraft could have arrived any sooner than they did. Specifically, we found the following:

- At approximately 12:30 a.m. on August 27—after it became apparent that firefighters would be unable to contain all the spot fires below the highway—the type 3 incident commander contacted an Angeles National Forest dispatcher to request that three air tankers be ordered to arrive at the fire at 7:00 a.m. At the time he made his request, the incident commander recognized that it might not be possible to get the air tankers at the fire by 7:00 a.m.[30]
- Following the incident commander's request for the tankers, another dispatcher entered the request at 12:55 a.m. into a computerized ordering system known as the resource ordering and status system (ROSS).[31] This dispatcher told us that before she formally submitted these requests to the geographic area coordination center,[32] which would identify available air tankers and assign them to the fire, she called the coordination center and was told not to formally submit the request until the coordination center knew which air tankers would be assigned. As a result, the orders remained in ROSS but were not formally submitted to the coordination center. The coordination center dispatcher, however, disputed this account, saying she does not recall this conversation and would never tell a dispatcher not to formally submit a request because it is important to keep track of all requests for air tankers so the center can document if it is unable to fill a request. The Department of Agriculture's Inspector General examined this issue and determined that records of phone calls between the Angeles National Forest and the coordination center relating to this matter do not exist, so it is not possible to determine whether the forest dispatcher called the coordination center or what discussion, if any, took place between them.[33]
- About 2 minutes after the incident commander's request (and about 20 minutes before the dispatcher entered the air tanker request into ROSS), two dispatch center officials discussed the incident commander's request for air tankers. According to a transcript of this conversation, one official said that the dispatch center did not need to request the tankers from the coordination center and would instead "divert" air tankers that were assigned to the Morris Fire.[34] According to this official, this conversation reflected his belief, based on an earlier discussion he had had with the coordination center, that the coordination center was unlikely to assign more air tankers to a second fire occurring at the same forest, but it did not reflect any decision on his part about whether the dispatcher should place the

order. He said that neither he nor the other dispatch center official involved in this conversation had any role in processing the request for the air tankers.

- Angeles National Forest fire management officials and dispatch center personnel told us they were aware that four air tankers had been tentatively assigned to the Morris Fire for August 27, and they ultimately decided to divert the needed air tankers from the Morris Fire to the Station Fire because the Station Fire had become a higher priority overnight. The manager on duty at the Fox Air Tanker Base in nearby Lancaster, where the planes were stationed, told us that a dispatcher informed him of this decision at about 6:30 a.m. on August 27. A fourth air tanker was requested for the Station Fire at 8:21 a.m.
- Subsequently, at about 8:40 a.m., an Angeles National Forest dispatcher who had recently come on duty was notified that the air tankers were preparing to take off from Lancaster for the Station Fire. The new dispatcher reported that she was not aware that the requests for these tankers had already been entered into ROSS and she entered requests for these same tankers into the system at that time.[35] She reported that she realized at about 9:00 a.m. that she had entered a duplicate order and, after confirming the orders were duplicates, she canceled the original orders in ROSS.
- Ultimately, the air tankers arrived over the Station Fire about 2 hours later than requested. The first tanker took off from Lancaster at 8:42 a.m., and the other tankers took off at 8:51, 8:56, and 9:03, according to incident records. Flight time from Lancaster to the Station Fire was about 10 minutes.

Regardless of how or when the air tankers were ordered, it appears unlikely that any federal air tankers could have arrived over the Station Fire sooner than they did.[36] Angeles National Forest and geographic area coordination center officials told us that the air tankers arrived later than requested by the incident commander because the crews were not allowed to report to work earlier because of mandatory rest requirements enacted to ensure crew safety, and no other federally contracted air tankers were located close enough to have arrived at the fire any earlier. On the night of August 26, 10 federal air tankers were located in Southern California. Five of them were in Lancaster, the location of the closest base to the Station Fire; four were in Fresno; and one was in San Bernardino. The crews for the five air tankers in Lancaster all went off duty at 8:00 p.m. on August 26 and were scheduled to come back on duty at 7:00 the next morning. The crews for four of the other five air tankers also went off duty at around 8:00 p.m.[37] Under applicable contract requirements and federal interagency fire aviation policies, crews must receive 10 hours of uninterrupted rest before their shift, meaning that the crews would have been eligible to report to work at 6:00 a.m., an hour before their scheduled arrival time.[38] If, however, an aircraft manager had called the crews before 6:00 a.m., the call would have been considered an interruption to their rest period, and the crews would have been required to wait 10 hours after receiving the phone call before they could have reported to work.[39] No federal air tankers were located in Northern California on the night of August 26; the next nearest federal air tankers were located in Ogden, Utah, and Pocatello, Idaho.

Once the air tanker crews reported for duty at the air tanker base at 7:00 a.m. on August 27, several standard preflight tasks had to be done before they could take off for the Station Fire. According to the tanker base manager on duty that morning, the crews first performed a preflight inspection of their aircraft, which took approximately 5 minutes. He reported the crews then received a briefing about both the Station Fire and the Morris Fire; this briefing lasted approximately 20 to 30 minutes, which is longer than a typical briefing, but the proximity of the two fires made it critical to discuss flight patterns to avoid collisions. The aircraft were then loaded with fuel and retardant.[40] Once these tasks were completed, the pilots warmed up the engines and did required preflight checks, which can take up to 30 minutes. The planes were ready to fly between 8:42 and 9:03 a.m.

Although it appears unlikely that any federal air tankers could have arrived at the Station Fire any earlier, some observers noted that the state firefighting agency, CAL FIRE, had air tankers that might have been available to respond more quickly. A CAL FIRE official told us that on the morning of August 27, the agency had three air tankers—located in Ramona and Hollister—that were not assigned to other fires and might have been available if the Forest Service had requested them for the Station Fire. The official told us that the crews for these tankers were not scheduled to report until 10:00 a.m., although he said that if the air tankers had been requested, the agency might have been able to bring the crews on duty earlier. According to this official, the crews for the air tankers in Ramona went off duty at 6:49 p.m. on August 26 and were eligible to return to duty at 4:49 a.m. on August 27.[41] The official estimated that if the tankers had been requested, they might have been able to arrive at the Station Fire at approximately 7:00 a.m., but he told us that because the request was never made, he does not know whether the air tankers would or would not have been available. Forest Service officials at the coordination center told us they did not request the air tankers from CAL FIRE because they did not receive a request for air tankers from the Angeles National Forest's dispatch center. The officials noted, however, that they did not believe CAL FIRE would have allowed the tankers to be used because the three air tankers in question were the last unassigned air tankers in California, and CAL FIRE would likely have retained them for initial attack, given the threat of new fires starting elsewhere in the state.

Irrespective of the controversy over tanker arrival times, some Forest Service fire management officials have suggested that the delay in the air tankers' arrival likely had little effect on firefighters' ability to contain the fire that morning. Officials noted that the section of the fire posing the greatest risk was the spot fire below the highway—the section that later in the morning spread down canyon and escaped—and the air tankers would most likely not have been used on this section. They would more likely have been used on ridgelines above the highway because, according to these officials, it would have been unsafe for air tankers to fly in the steep canyon below the highway. In addition, the tankers would have had to fly high enough to clear the power lines crossing the canyon, and drops from such heights would likely have been ineffective because the retardant would have dispersed too much to penetrate the vegetation canopy.

Adequacy of Very Large Air Tanker Use

Some observers questioned why the Forest Service did not use very large air tankers more often in fighting the fire. On the basis of our review, we determined that in some instances the use of very large air tankers was suggested, but they were not ordered, and in other cases when these aircraft were ordered, they were not available or the order was subsequently canceled. For example, on August 27, the use of a very large air tanker was suggested by the air tactical group supervisor—the official in charge of coordinating aircraft for the Station Fire—but the tankers were not ordered by the incident commander.[42] The air supervisor reported that he requested a very large air tanker three times that day, but each request was denied. According to the air supervisor, he believed the aircraft would be useful because of the length of fireline needing to be constructed. Transcripts of radio communications indicate that the incident commander trainee for August 27 told a dispatcher that the incident would not be ordering the very large air tanker requested by the air supervisor.[43] The incident commander trainee told us that he discussed the air supervisor's requests with the incident commander and with an Angeles National Forest fire management official and that, on the basis of information they received from firefighters in the field, the incident commander determined that, because of rough terrain and the presence of power lines in the area, there was no suitable location for the very large air tanker to make a safe drop, and he therefore did not order the aircraft that day.

Our review of incident records also found instances where orders for very large air tankers could not be filled because of mechanical issues or were canceled for other reasons. For example, officials noted that they requested the Martin Mars on August 29, one of the key days of the fire, but the aircraft was not available that day because of mechanical issues.

Incident records also indicate that a DC-10, 747, and Martin Mars were ordered by the incident management team shortly before 6:00 p.m. on September 2, to be used the following day, but that the orders were canceled shortly after 7:00 p.m. on September 3. The Forest Service provided no documentation that would explain why these orders were canceled. However, Angeles National Forest officials told us that the orders were likely canceled because incident officials had requested the air tankers on September 2 in anticipation of needing them on September 3 but then did not request the tankers for any missions on that day. The officials said that orders would have been canceled in this situation to keep the dispatch ordering system up-to-date.

Nevertheless, incident management and Angeles National Forest officials told us they believed that very large air tankers were used in the appropriate situations, such as protecting communities along the foothills, and that using them more frequently would not have been effective.[44] According to Forest Service and incident management officials, very large air tankers can be useful tools for firefighting because they can carry more retardant than other aircraft, although their size limits their use in rough terrain.[45] In all, beginning on August 29, the DC-1 0 was used on at least 8 days, and the 747 and Martin Mars were each used on at least 2 days, according to incident documents. Forest Service direction requires the forest supervisor and regional fire management officials to approve the use of very large air tankers, and incident management team members told us that all of their requests to use the air tankers were approved.

Timeliness of Ordering the Incident Management Team

Some observers questioned why the Forest Service did not order an incident management team as soon as it became apparent that the fire would not be contained by initial attack, suggesting that forest officials should have acted earlier to order either a type 1 team or a less qualified but more easily mobilized type 2 team, which could have arrived faster. The observers told us that such actions might have allowed firefighters to better prepare for the extreme fire behavior of August 29 and 30. On the basis of our review, we determined that the Angeles National Forest ordered a type 1 team at approximately 2:00 p.m. on August 27. It typically takes 18 to 24 hours to mobilize a type 1 team, according to officials, and the team assumed command of the incident at 2:00 p.m. on August 28—approximately 24 hours after being ordered and nearly 2 days after the fire began. For several hours before Angeles National Forest officials ordered the type 1 team, however, the fire was burning in ways that indicated that either a type 1 or type 2 team might be needed to manage it. For example, a complexity analysis performed by the incident commander during the early morning hours of August 27 concluded that the fire's intensity, the number of firefighting assets being used, and the planning and logistical actions needed to support firefighting efforts all indicated the need for a type 1 or type 2 team. Moreover, firefighters assigned to the fire on the morning of August 27 reported that fire behavior was extreme as early as 8:30 a.m. and that firefighters had to retreat several times because the fire was so intense—further indication that a type 1 or type 2 team could have been needed. If Angeles National Forest officials had ordered a type 1 team at the first indication that it could have been warranted, it is possible the team could have assumed command at 6:00 a.m. on August 28, 8 hours earlier than it ultimately did.

In addition, some observers have stated that Angeles National Forest officials should have ordered a type 2 team to manage the fire until the type 1 team arrived. It typically takes 4 to 6 hours to mobilize a type 2 team, according to officials—substantially less time than required to mobilize a type 1 team. Angeles National Forest officials told us they considered ordering a type 2 team but decided against it because they did not want to have two command transitions in a short time—especially given that such transitions are generally considered to temporarily increase risk to firefighters because of changes in responsibilities and lines of communication. Instead, officials decided to make the Station Fire a "branch" of the nearby Morris Fire, so that the type 2 team managing the Morris Fire could provide planning and logistical support for the Station Fire until the type 1 team could arrive. The incident commander of the type 2 incident management team told us this arrangement was workable but led to some difficulties as well. For example, he said the command post for the Morris Fire was approximately 20 miles away from the Station Fire, and it was difficult to keep abreast of developments at both fires, particularly as conditions continued to deteriorate at the Station Fire. A senior member of the type 1 team told us that earlier mobilization of either a type 1 or type 2 team might have allowed incident managers to better plan firefighting strategies and tactics but also said that given the fire's extreme behavior, he did not know whether earlier mobilize-tion would have made a difference in the outcome. Some observers, however, have stated that earlier mobilization of a team might have allowed firefighters to take more action before the fire began its period of rapid growth on August 29.

Ordering Forest Service Assets Rather Than Other, Potentially Closer, Assets

Some observers questioned whether, in response to a memorandum from the regional office overseeing the Angeles National Forest, the agency sometimes mobilized Forest Service firefighting assets rather than local ones—even though the Forest Service assets were farther away and would take longer to arrive—and whether this decision reduced the effectiveness of the agency's response. Shortly before the Station Fire, the Forest Service's Pacific Southwest regional office had issued a memorandum stating that because of the potential for a budgetary shortfall, dispatch and incident management teams should, where appropriate, be directed to order Forest Service firefighting assets when responding to a fire—rather than use contract or state and local agency assets—and, if non-Forest Service assets are used, to replace them with Forest Service assets as quickly as possible. On the basis of our review, the role this memorandum played in the ordering of assets for the Station Fire is unclear. Officials at the Angeles National Forest and the interagency geographic area coordination center said that the memorandum did not affect their decisions about which assets to order, but incident management team members suggested otherwise. Available documentation indicates that assets from multiple ownerships, including both the Forest Service and local jurisdictions, were ordered, but the documentation did not include the information needed for us to determine the extent to which some firefighting assets may have arrived later than requested because Forest Service assets were ordered instead of available nonfederal assets. Specifically, we found the following:

- *Dispatch officials and incident management team members provided conflicting accounts of the memorandum's role.* Angeles National Forest dispatch personnel and coordination center officials told us the memorandum did not affect how they ordered firefighting assets for the Station Fire. Dispatch personnel said, for example, that during the fire's early days, they ordered the closest available assets regardless of which agency owned them—a practice consistent with the California Interagency Mobilization Guide, which directs how several wildland firefighting agencies, including the Forest Service, are to order firefighting assets in California. A dispatch center official told us that later in the fire, dispatchers would have ordered Forest Service assets first, provided that the assets could have been expected to arrive when requested by the incident management team. The official explained that if such assets could not be mobilized by the time requested, dispatchers could still order nonagency assets that might be able to arrive more quickly. According to members of the incident management team, however, some of the firefighting assets they requested were slow to arrive, particularly during the first few days after the team assumed command of the incident on August 28. These team members told us that the delay might have resulted partly from dispatchers ordering Forest Service assets rather than closer, local assets, but they also said that the delays ultimately had little effect on their ability to fight the fire. One incident management team member told us he talked to an Angeles National Forest dispatcher about the delay and was told by the dispatcher that because of the memorandum, dispatchers were ordering Forest Service assets, which were coming from out of the area.
- *Limitations in the available documentation on orders for firefighting assets prevent a conclusive determination on the memorandum's effect.* From the information available, we could not conclusively determine the extent to which arrival of assets may have been delayed because agency assets were ordered instead of assets available locally, although our review of the data provides some information regarding this issue.[46] Data on orders for firefighting assets are maintained in ROSS, and although the data in this system were sufficient for us to

analyze the ownership of each asset, they did not identify assets that may have been available to respond to the fire but were not ordered. As a result, we were unable to determine the availability of nonagency assets located closer to the Station Fire in instances where the Forest Service ordered an agency asset that arrived later than requested. We reviewed data on requests for two types of assets that incident management team members said were slow to arrive: type 3 fire engines[47] and personnel to fill incident management team positions known as operations branch directors. For type 3 engines, of the 68 engines assigned on August 27 and 28, more than 90 percent were non-Forest Service engines, suggesting that the Forest Service did not consider agency ownership when ordering assets. On August 29 and August 30, however, almost half the orders for type 3 engines were filled by Forest Service engines, many of which came from outside the area. Forest Service officials, as well as a local agency official, told us that local fire departments have a limited number of type 3 engines, and the more such engines the Station Fire needed as the fire progressed, the more likely it would have become that assets would have had to be ordered from farther away. From August 27 to August 30, ROSS records show that dispatchers were having difficulty filling orders for type 3 engines. For example, on the afternoon of August 29, a dispatcher entered a notation in ROSS that no more strike teams of type 3 engines were available in California and that dispatchers would need to place orders for single type 3 engines through the National Interagency Coordination Center.

For operations branch directors, we were unable to determine the extent to which personnel arrived later than requested. ROSS records show that for the 14 orders filled on August 27 through August 30, 9 were filled with Forest Service personnel, some of whom had to travel long distances, and 4 of the 5 non-Forest Service operations branch directors were assigned on the evening of August 29. A senior federal and nonfederal official both told us that many nonfederal personnel qualified as operations branch directors in Southern California should have been available to mobilize more quickly. Records in ROSS do not identify assets that were available but not ordered, however, so we were unable to determine the availability of nonfederal personnel during this time frame.

Forest Service officials from the Pacific Southwest regional office (which issued the 2009 memorandum), the geographic area coordination center, and the Angeles National Forest all told us they do not believe the memorandum resulted in delays in filling requests for assets to respond to the Station Fire. These officials said that the memorandum stated that the Forest Service should use agency assets "as appropriate" and did not change or repeal existing Forest Service policies and procedures for ordering firefighting assets. Under the policies and procedures in effect at the time of the Station Fire, agency officials were directed to use the closest available resource to respond during a fire's early stages. Furthermore, the officials also said that the memorandum stated that in cases where dispatchers believed it was appropriate to order agency assets first, the ROSS order should indicate that Forest Service assets were being specifically requested. Our review of ROSS data for August 27 through August 30 showed that 22 of 164 orders for type 3 engines requested a specific engine, but none of these were Forest Service engines. For operations branch directors, 6 of 14 orders specifically requested a particular person, two of whom were from the Forest Service. A Forest Service official told us that one of the people requested was already a member of the incident management team and the other was an official from a nearby national forest.

Adequacy of Efforts to Protect Homes in Big Tujunga Canyon

Some observers questioned whether more action could have been taken to protect the homes in Big Tujunga Canyon and, more fundamentally, whether the Forest Service was even aware that homes in the canyon were at risk. For example, some residents and others have questioned why, once the fire entered the upper reaches of Big Tujunga Canyon, firefighters did not use air tankers to drop retardant or take other actions to suppress or contain the fire before it burned into the populated areas farther down the canyon or to mitigate its intensity as it burned through the canyon. Our review determined that by the afternoon of August 28, the fire had burned over a ridge into the upper reaches of one of Big Tujunga Canyon's side canyons. According to firefighters and some residents we spoke with, however, the fire was not burning at high intensity or spreading very quickly. Some residents told us that if retardant had been dropped on portions of the side canyon early enough, the fire might have been kept out of the main canyon entirely, or the intensity of the fire might have lessened as it burned through, perhaps to the extent that the plume-dominated conditions, which made the fire in the canyon so intense on August

29, would have been unable to develop. Similarly, some residents told us that firefighters could have dropped retardant closer to homes in the canyon on either August 28 or 29, so that the fire would have been less intense, which may have allowed the firefighters who were in the canyon to safely take additional steps to protect homes. One resident said that he believed these actions would have helped save homes in Big Tujunga Canyon, but he believed they were not taken because firefighters were focused on protecting larger foothill communities such as La Cañada Flintridge and La Crescenta. This resident also noted that while such a decision would have been understandable, he still believed that firefighters could have taken more action to protect the homes in Big Tujunga Canyon. Another resident told us that if firefighters had not planned to take more action, they should have notified residents sooner so residents could have packed some of their possessions and evacuated earlier.

Incident records note the use of air tankers to drop retardant on August 28 but do not specifically note the location of the retardant drops. Two firefighting officials told us that air tankers dropped retardant on the ridgelines along Lukens Ridge above Big Tujunga Canyon on August 28 in an effort to keep the fire from burning farther into the canyon; officials told us they did not make drops farther to the east (near Hoyt Mountain), an area where the fire had burned into the upper reaches of Big Tujunga Canyon's side canyons. One of the officials said that he did not believe retardant drops in this area would have been effective in part because the vegetation was too dense for the retardant to be able to reach the ground (which would have allowed the fire to burn through vegetation close to the ground) and another official told us that the area was too steep to safely place firefighters to build a fireline once the retardant was dropped. Agency officials noted that the purpose of retardant is not to put out a fire, but to slow it down and lower its intensity so that firefighters can work to establish a fireline to contain the fire. Officials told us that the incident management team took other actions to try to protect the homes in the canyon. For example, incident management officials said they used bulldozers on August 28 to try to build a fireline across a portion of the canyon, from the top of nearby Lukens Ridge down toward the canyon bottom. This line was built down steep slopes, the officials said, and even though the bulldozer operators were very experienced, they were unable to create a line all the way to the bottom of the canyon. The officials said they knew it was unlikely they would be able to complete this line, but they thought it was worth the effort.[48] Because the bulldozers were unable to reach the canyon bottom, and the remainder of the line would have had to be built by hand—a task one Forest Service official estimated would have taken several days—the fireline was not completed. Two experienced firefighters told us, however, that even had the line been completed, they believed the fire behavior in the canyon on August 29 was too extreme for the line to have held. Officials told us they did not drop retardant in the canyon on August 29—the day the fire burned through the canyon—because there was too much smoke in the morning for air tankers to safely drop retardant, which was the result of the temperature inversion. Once the inversion lifted and the smoke cleared, the fire quickly developed plume-dominated conditions, which again prevented tankers from flying over the area.

Some observers also questioned why more fire engines were not assigned to Big Tujunga Canyon and why firefighters who were in the canyon did not take more actions to protect homes before the fire burned into the canyon. These observers suggested that had more fire engines been in the canyon or had more actions been taken, fewer homes might have burned. From our review, we determined that three strike teams, each with five fire engines, were assigned to the canyon on August 29.[49] According to incident management officials and firefighters assigned to the canyon whom we spoke with, having additional engines may have helped them protect homes, but additional engines would not have had enough places to safely take shelter during extreme fire behavior. It would therefore not have been safe to have more engines in the canyon. They also stated that many residences in the canyon lacked sufficient defensible space. A strike team leader assigned to the canyon told us firefighters worked to create defensible space by removing vegetation and other flammable materials from around residences before the fire burned through.[50] Firefighters said they also spent time talking with residents who did not want to evacuate to convince them of the importance of evacuating, which reduced the time they had to prepare structures before the fire arrived. As it became apparent that the fire was going to burn through the area sooner than expected, one of the strike team leaders said they also hosed down homes and applied a foam retardant to a Forest Service ranger station for additional protection. Firefighters said they continued these efforts until the fire forced them to retreat to safety zones. After the fire passed, firefighters said they were able to leave the safety zones, and they resumed their efforts to protect homes that had not been destroyed—and successfully saved several homes—but, they said, their first priority became providing medical assistance to civilians who were hurt during the fire. In general, incident management officials, as well as firefighters assigned to the canyon, told us that firefighters can do little to protect homes during extreme fire behavior like that exhibited in Big Tujunga Canyon on August 29.

Some Big Tujunga Canyon residents raised a still more fundamental issue, stating that firefighters were not fully aware of the number or locations of homes in the canyon. For example, one resident noted that even though dozens of canyon homes were destroyed, the Forest Service reported only three homes destroyed—suggesting that the incident management team did not know how many homes were in the canyon.[51] During our review, members of the incident management team and Angeles National Forest officials pointed to the creation, on August 28, of a structure protection group specific to Big Tujunga Canyon—and the presence of fire engines and crews when the fire burned through on August 29—as indications that the incident management team was aware of, and responsive to, the danger to those structures. The supervisor responsible for this structure protection group told us he drove through the canyon on the evening of August 28 and met with local forest officials so that he could get acquainted with the area and better understand the number and location of homes and other structures, water sources, access points, and other information. The supervisor said that he was also given maps identifying residences in the canyon, including homes on private property and homes on land leased from the Forest Service. Regarding the inaccurate information in incident documents about the number of homes destroyed in the canyon, the structure protection group supervisor told us that although he did not know the exact number of homes destroyed by the fire at the time the reports were prepared, he knew that many homes had been destroyed. However, incident management officials wanted to wait to update the early numbers cited in incident documents until complete and accurate information about the total number of destroyed homes could be established. The supervisor said that because developing this information took time, some observers may have interpreted the incorrect information as evidence that incident management officials were not aware that a larger number of homes had in fact been destroyed, which, he said, was not the case.

Timeliness of Evacuating Big Tujunga Canyon

Some residents of Big Tujunga Canyon expressed concern that they were not notified early enough of the need to evacuate, while others said that they never received notice of a mandatory evacuation. Neither the Los Angeles County Sheriff's Department—the agency in charge of implementing the evacuation—nor the Forest Service could provide us with conclusive documentation of when the canyon was evacuated.[52] Members of the incident management team told us they believed their initial plan for evacuating the canyon on August 29, were it to become necessary, was sufficient but that the unexpectedly rapid advance of the fire that day contributed to their inability to provide more notice for evacuating canyon residents.

Many residents of Big Tujunga Canyon whom we interviewed expressed concern about how the canyon was evacuated. Some residents said they should have been given more notice, with one saying evacuations should have begun as early as the morning of August 28. Some residents told us they were never notified of the mandatory evacuation. One resident said that neither she nor some of her neighbors were notified, and another said that a sheriff's deputy told her at approximately 9:30 a.m. on August 29 that a voluntary evacuation was in place but that no one notified her of a mandatory evacuation. She said she evacuated her home at around 11:30 a.m., shortly before the fire arrived. One resident told us that residents were not notified by means such as loudspeakers or door-to-door contacts. Another resident told us that she heard vehicle loudspeakers notifying residents to evacuate but that some residents were unable to hear the announcement because they were inside their homes, a particular concern for elderly or disabled residents who would have had greater difficulty evacuating quickly.

Our review found that the events surrounding the evacuation of Big Tujunga Canyon were not fully described by available documentation, and officials' accounts of these events were not precise.[53] According to transcripts of radio communications, around noon on August 28, Angeles National Forest and incident management officials discussed the possibility of evacuating Big Tujunga Canyon, and some observers we interviewed told us that a voluntary evacuation was in place by Friday evening. Records from the Sheriff's Department indicate that shortly before 5:00 p.m. on August 28, the department directed its personnel to begin notifying residents of Big Tujunga Canyon to be prepared to evacuate. Department records also indicate that at 8:52 a.m. on August 29, the department directed its personnel in the canyon "to conduct evacuations on order of the fire department." A senior Sheriff's Department official told us that this record indicates that department personnel were ready to evacuate residents once they received notification from incident management team officials. According to incident management officials, a mandatory evacuation was ordered at approximately 10:20 a.m. on August 29, shortly after the fire reached the Grizzly Flats plantation, the evacuation "trigger point" identified the previous evening by the supervisor of the structure protection group. A senior Sheriff's Department official told us that the department had many personnel in the canyon on August 28 and 29. The department, however, could not provide documentation about the actions its personnel took to carry out the evacuation, and this official told us that events of that day may not have been fully documented because the situation changed so quickly. Firefighters and Angeles National Forest officials present in the canyon told us that they also notified some residents to evacuate, but that some residents they contacted told them they intended to stay.

The structure protection group supervisor told us he believed that the Grizzly Flats trigger point would have allowed enough time to evacuate the canyon, an assessment reached in consultation with an Angeles National Forest official familiar with the area. Once the fire reached the trigger point, however, it burned through residential areas of Big Tujunga Canyon in only a few hours, much faster than they had expected. Incident documents indicate that on August 29 the team recognized that extreme fire behavior could occur in this area, which could cause the fire to spread rapidly, but members of the team told us there was no indication on August 28 or the morning of August 29 that the fire would move through the canyon as quickly as it did.[54]

Adequacy and Appropriateness of Firefighting Strategies and Tactics

Some observers raised questions about various aspects of the Forest Service's strategic and tactical approach to fighting the Station Fire, suggesting that a more aggressive response may have allowed the agency to contain the fire sooner. We heard a wide range of opinions on the adequacy and appropriateness of the Forest Service's response, but we had no definitive way to determine whether or to what extent the ideas we heard would have made a difference in controlling the fire's spread or reducing the extent of its damage. Incident management officials told us they considered numerous factors—including the risk to firefighters and the public, the structures and resources threatened by the fire, and the likelihood that particular firefighting actions would be successful—in determining the strategies and tactics to use to respond to the fire. Of these factors, Forest Service policy states that the first priority is the safety of firefighters, other personnel, and the public, and, officials told us, this policy guided the decisions made in fighting the fire.

Questions raised about the adequacy and appropriateness of firefighting strategies and tactics included the following:

- Some observers questioned why firefighters decided to let the fire burn into the green island—the source of the embers that started the spot fires below the highway on the first night. The green island was on a steep vegetated slope that presented significant firefighting challenges. The incident commander and other firefighters at the scene told us they recognized the potential risk of embers from the green island and tried several actions to reduce that risk, all of which proved unsuccessful. For example, firefighters first tried to attack the fire's left flank directly, which, if successful, would have left no unburned fuel inside the firelines; the fire's intensity, however, prevented firefighters from safely completing this action. Firefighters told us they then attempted to place a hose line above the green island, so they could get water more easily into the unburned area, but the steep terrain and the fire's intensity prevented safe completion of this action as well. Firefighters also told us they tried using deck- mounted water guns on their fire engines, known as "monitors," to shoot water up into the fire and reduce its intensity, but the fire was burning too far up the hillside for the water guns to have much effect. Firefighters said they then attempted to set fire to the green island itself, so it would burn more quickly and consume all the fuel in the area before down-canyon winds began to blow, but this attempt too was unsuccessful.
- Some observers said that if firefighters had directly attacked all the spot fires that started below the road on the first night, they might have been able to contain the fire. According to Forest Service officials responsible for the fire response that night, however, the spot fires were burning on very steep slopes, and it was not safe for firefighters to attack some of those fires at night. Angeles National Forest officials and firefighters told us they were especially mindful of safety issues because of the history of firefighter fatalities in the Angeles National Forest in fire conditions similar to the Station Fire's. However, some Forest Service officials we interviewed—including some involved in making decisions about how to fight the fire—also suggested that the aggressiveness of the agency's response was limited by the experience of the firefighters available to respond, in part because some firefighters familiar with the area were already assigned to the nearby Morris Fire. These officials told us that had more experienced firefighters been at the Station Fire the first night, they might have been able to safely attack more of the spot fires.[55]

- Some observers noted that once the Station Fire began to spread quickly, the incident management team attacked the fire more aggressively along the foothills where it threatened populated areas but adopted indirect firefighting strategies in other areas, suggesting that the team deliberately allowed the fire to burn farther into the national forest and, in particular, into wilderness areas. The incident commander and members of the incident management team told us their goal was to fully suppress the fire in all areas, which was consistent with Angeles National Forest policy, but said that the fire's speed and intensity were often too great for safe direct attack. Several firefighters we interviewed, including some with many years of firefighting experience, stated that the fire's behavior on August 29 and 30 was more extreme than they had ever seen in their careers. On August 29 and 30 alone, the fire burned approximately 90,000 acres as it burned many miles in every direction; under these conditions, the incident commander determined that indirect strategies were to be used, a decision the incident commander said was agreed to by the other agencies in the unified command and by Angeles National Forest officials.
- Other observers told us they recognized that extreme fire behavior limited the available opportunities for direct attack but did not eliminate such opportunities entirely; these observers told us that the incident management team missed opportunities where more- aggressive action could have been safely taken. For example, some observers—including officials assigned to the fire—suggested that as the fire was burning north, firefighters could have tried to establish more firelines on ridgelines within the forest boundary, rather than wait for the fire to burn closer to communities along the forest's northern boundary. Similarly, one observer noted that on the fire's western flank, many ridgelines and a substantial network of roads might have been incorporated into backfiring operations to help contain the fire. Furthermore, some observers noted that the speed of the Station Fire often meant that it burned through places before planned tactics, such as building a fireline, could be completed. The observers said the incident management team should have been executing several contingency plans at the same time, so if the fire overtook one set of containment lines before they could be completed, firefighters would already be working on other lines farther away. Members of the incident management team noted that they established

a contingency planning group to identify such opportunities and told us they constructed contingency lines where they believed those lines could be successful in slowing the fire, but the fire's speed and size limited opportunities to effectively engage the fire or implement some of their contingency plans.

- Some observers noted that the Forest Service and Los Angeles County had not developed a plan for how best to defend the communications infrastructure on Mt. Wilson and the surrounding area in the event of a wildland fire and that when the Station Fire began to threaten Mt. Wilson, the Forest Service and the county initially disagreed over how best to respond. Incident management team and Los Angeles County officials told us that the county believed a greater firefighter presence was needed on Mt. Wilson but that the Forest Service did not believe that such a presence would be safe or effective. In addition, Forest Service officials told us that previous wildland fires in the area burned themselves out before reaching the communications facilities atop the mountain. Ultimately, the Forest Service agreed to have firefighters create defensible space around the facilities and have helicopters drop a substantial amount of water and retardant to help protect the area.

Although some observers criticized the efforts to suppress the Station Fire, many others we spoke with commended the Forest Service's response. Observers from multiple agencies we interviewed said that firefighters faced extremely difficult and dangerous conditions and that the decisions made and actions taken helped protect thousands of homes and the safety of the thousands of firefighters involved in fighting the fire. In particular, some firefighters pointed to the actions taken to keep the Station Fire from burning into La Cañada Flintridge and La Crescenta, and from burning the important communications facilities on Mt. Wilson, as key successes in the response.

Sufficiency and Capability of Aviation Assets Agencywide

Some observers also raised a larger concern about whether sufficient aviation assets were available to respond to the Station Fire and whether the response was indicative of a broader need for more, or different, assets to respond to the damaging wildland fires that occur in Southern California and elsewhere around the country. Specifically, noting the delayed arrival of air tankers, some observers questioned whether the Forest Service and other federal firefighting agencies have enough air tankers to respond to potentially damaging fires, particularly given the reduction in the number of air tankers available under federal contract— from 44 in 2002, to 19 in 2009, to 14 in August 2011. One observer also suggested that in light of the growing number of large, damaging wildland fires over the past decade, federal firefighting agencies should take steps to reduce the damage such fires cause, including acquiring more very large air tankers; adopting technology enabling aircraft to fight fires at night; and improving aircraft capability to drop retardant in high winds, which frequently occur in Southern California.

The Forest Service is evaluating some of these concerns but has not yet determined if it needs additional aviation assets or assets with different capabilities. Specifically, the Forest Service is conducting three studies examining different aspects of its aviation program, which agency officials expect to complete in late 2011 or in 2012. One study, responding to concerns raised by the Station Fire, is examining the benefits and risks of using aircraft to fight fires at night. Another is examining the agency's entire aviation program to determine the appropriate mix of air tankers and helicopters. A third study is evaluating how the Forest Service should acquire new air tankers for its fleet, examining, among other things, different aircraft models and whether the agency should purchase or lease them or contract with private companies.

Lessons from the Station Fire Have the Potential to Improve Future Wildland Fire Response

Because of its magnitude and impact, the Station Fire offers several important lessons that, if applied, could help improve wildland fire response in the future, including in Southern California. Through our reviews of incident documents and agency reports, interviews with Station Fire officials and others, and our previous work on wildland fire management, we identified a variety of areas where lessons can be learned from the Station Fire and applied to future firefighting efforts. These lessons include the importance of (1) determining the appropriate role of night-flying aircraft, (2) having transparent processes for ordering and mobilizing firefighting assets, (3) tracking aircraft water and retardant deliveries, (4) predicting fire behavior under a variety of conditions, and (5) having systematic methods to identify needed firefighting assets. Some of these lessons have also been discussed in a "lessons learned" report the Forest Service prepared in October 2010 on the basis of its experience with the Station Fire, in which the agency has described some of the steps it will take to implement the lessons.[56]

Importance of Determining the Appropriate Role of Night-Flying Aircraft

The Station Fire exemplifies a situation where using night-flying aircraft may allow the Forest Service to suppress a fire before it escapes efforts to contain it. The Forest Service, however, does not own or operate night- flying helicopters, although it allows the use in national forests of such helicopters operated by nonfederal firefighting agencies. For example, in Southern California, five counties and two cities, including Los Angeles County and the city of Los Angeles, have helicopters that are capable of flying firefighting missions at night, according to Forest Service officials.[57] In response to concerns raised about the Forest Service's not using night- flying aircraft on the Station Fire, the agency has made two changes affecting night-flying operations above national forests in California. First, in October 2010, the Forest Service changed its night-flying policy to authorize the use of turbine-powered single-engine night-flying aircraft, including helicopters, which had previously been prohibited.[58] This change increased the number, from three to seven, of nonfederal agencies allowed to operate night-flying helicopters over national forest lands in Southern California. Second, the Angeles National Forest and Los Angeles County changed their cooperative firefighting agreement to clarify the circumstances under which the county's night-flying helicopters may be used to fight fires in the Angeles National Forest. The 2011 agreement states that, "When appropriate, Incident Commanders can approve the use of helicopters at night when the fire is an immediate threat or may become a threat to life and/or property. This use of helicopters for aerial night fire fighting applies to all Angeles National Forest lands."[59] The agreement in place at the time of the Station Fire stated that, "District helicopters may be dispatched and utilized on fires during darkness on Federal lands when supporting the protection of life or property (under Unified Command)."[60] According to some Forest Service and Los Angeles County officials, under the previous agreement it was unclear whether county night-flying helicopters could be used when a fire did not pose an immediate threat to life or property or if the fire was not burning close to county lands.

Following the Station Fire, the Forest Service also began a nationwide assessment of the agency's night-flying operations and established a National Night Flying Operations Working Group. The agency is studying various aspects of nighttime helicopter use, including risk, effectiveness, and availability of night-flying helicopters, and is evaluating whether it should develop its own night-flying capability. A senior Forest Service official told us in May 2011 that he expected the study to be completed by August 2011, but the study had not been completed at the time of our review. This official told us that he expected the study's helicopter portion to be completed later in 2011. Officials attributed the delays to the complexity of the issues, including the complexity of ensuring the safety and effectiveness of night-flying operations and the need for a comprehensive cost analysis. As part of the Forest Service's efforts to study the issue, the agency contracted for a risk analysis that evaluated the available technology and reviewed past and current uses of night- flying helicopters.[61] This study identified 79 hazards—such as a pilot's reduced vision at night, aircraft lighting that is not certified for night operations, and inability to identify intended targets on the ground—that the agency believes it would need to mitigate before a night-flying program could be implemented; the study also identified 130 associated mitigation measures. The agency is developing an implementation strategy for addressing these hazards.

Importance of Having Transparent Processes for Ordering and Mobilizing Firefighting Assets

The Station Fire demonstrates the importance of having transparent processes for ordering firefighting assets to minimize miscommunication and improve record keeping. Following the Station Fire, the Angeles National Forest made several changes to its asset-ordering and mobilizing procedures, including (1) prohibiting informal requests for assets, instead directing that all requests be entered into ROSS (the centralized resource tracking system) and placed to cooperators or to the regional dispatch center, thereby clearly documenting which requests are filled and declined and by whom; (2) directing dispatchers to document all delays in obtaining ordered assets and relay this information to incident commanders so they may amend or alter plans of attack; and (3) documenting the time that individual assets are released from an incident and the reason for releasing them (e.g., because the agency providing the assets requested their release), thereby providing a record of the use of those assets and the reason for their reassignment. Forest officials told us these changes were implemented in February 2010. The Angeles National Forest revised its dispatch center operating guide in September 2011, directing dispatch personnel to enter all requests for assets into ROSS, and to inform incident commanders of any delays in obtaining ordered assets. Angeles National Forest officials told us they have also directed forest dispatchers to document any delays in obtaining ordered assets, the time assets are released, and the reason for the release, but they said this change was not included in the revised guidance. The controversy over how firefighting assets were ordered and mobilized for the Station Fire suggests that the Forest Service may benefit from providing similar direction to its dispatch centers agencywide. The Forest Service's "lessons learned" document identified potential lessons related to standardizing the agency's asset-ordering processes but did not identify specific steps the agency plans to take.

The Station Fire also demonstrates the importance of having clear guidance about which assets are to be ordered first. The controversy we identified over the effect of the Forest Service's August 2009 memorandum about ordering Forest Service assets first for budgetary reasons even though other, non-Forest Service assets might be closer suggests that the agency would benefit from clearly articulating the circumstances under which it is appropriate to order its own assets instead of state or local agencies' assets. Forest Service officials told us the memorandum was in effect only for 2009 and pointed out that it directed that agency assets be ordered first only "as appropriate." However, the memorandum did not specify the circumstances under which Forest Service assets should be ordered even if other, nonagency assets are closer, raising questions about the agency's overall policy in this regard. Consequently, additional clarification about these circumstances could assist fire officials in making decisions about which assets to order and could help nonagency observers, including the public, better understand the agency's asset-ordering policy.

Importance of Tracking Aircraft Water and Retardant Deliveries

The Station Fire demonstrates the importance of being able to track and identify the locations where aircraft deliver water and retardant. Because the Forest Service did not have this capability during the Station Fire, it had limited ability to identify the locations where aircraft had dropped water and retardant, which made it difficult for the agency to answer questions about its response to the fire. Recognizing the importance of having this capability, a senior Forest Service official told us that, by the end of 2012, the agency expects to equip all air tankers with a system to track retardant drops. The official also said that the agency is evaluating other possible improvements to its ability to track and evaluate water and retardant drops. For example, the agency is evaluating improvements that could allow it to evaluate the effectiveness of each drop by means of infrared imaging technology, which would assess changes in fire intensity or rate or direction of spread. Such information, if provided in real time, could help incident management officials determine not only which drops were effective but also which drops were not achieving the desired effect and should be reconsidered. Combining this type of information with drop costs could help the agency determine the cost-effectiveness of drops in different areas or by different types of aircraft—information that could in turn help the agency evaluate which types of drops and aircraft are best suited to different conditions.

Importance of Predicting Fire Behavior under a Variety of Conditions

The Station Fire's intensity and speed of spread also demonstrate the importance of having effective fire behavior models available to incident command officials who are making decisions about response tactics, strategies, and assets. Firefighters and incident management officials we interviewed said that even though they recognized that weather and fuel conditions were conducive to a high-intensity, fast-spreading fire, the ability of the Forest Service's primary tools for predicting fire activity— including modeling software known as FSPro, BehavePLUS, and FARSITE—are limited under certain conditions, particularly the plume- dominated conditions of key days during the Station Fire. Improving fire behavior modeling tools might allow incident managers and firefighters to better predict fire behavior and adjust their strategies and tactics accordingly. The Forest Service reported it is taking steps to improve the information used to predict weather conditions, which can greatly influence a fire's behavior, and that it is exploring how to better model fire behavior as well.

Importance of Having Systematic Methods to Identify Needed Firefighting Assets

The Station Fire also demonstrates the significance of issues we have identified over the last decade regarding the importance of having systematic methods of identifying needed firefighting assets. For example, we have previously reported that since 2002, the Forest Service, in conjunction with agencies in the Department of the Interior, has been developing a system known as fire program analysis (FPA), which is intended to, among other things, allow the agencies to identify the most cost-effective mix of fire-fighting assets—information that could help the agency determine the assets it needs and where best to locate them. Nevertheless, we reported in March 2011 that, after almost a decade of work, FPA's development continued to be characterized by delays and revisions, with the agencies several years behind their initially projected timeline for using it to help develop their budget requests.[62] The agencies are continuing to take steps to improve FPA, including submitting it to an external peer review. According to agency officials, the agencies expect to begin using FPA to inform their fiscal year 2013 budget requests. These efforts are still under way, and we have not evaluated the extent to which they may address the shortcomings we identified.

More fundamentally, in a series of reports dating to 1999, we have recommended that the Forest Service and Interior agencies develop a cohesive wildland fire strategy that identifies potential long-term options for reducing fuels and responding to fires, as well as the funding requirements associated with the various options.[63] We have reported that, by laying out various potential approaches, their estimated costs, and the accompanying trade-offs, such a strategy would help Congress and the agencies make informed decisions about effective and affordable long-term approaches to addressing the nation's wildland fire problems. Congress echoed our call for a cohesive strategy in the Federal Land Assistance, Management, and Enhancement Act of 2009, which requires the agencies to produce a cohesive strategy consistent with our recommendations.[64] In March 2011, the Forest Service and Interior released *A National Cohesive Wildland Fire Management Strategy*, which describes the three-phase process the agencies plan to use to develop a cohesive strategy. [65] As of December 2011, the agencies were working on the second phase of their efforts— developing regional assessments of, and strategies for addressing, wildland fire risk—and had developed a draft document titled *A National Cohesive Wildland Fire Management Strategy, Phase II National Report*. However, neither the phase I nor the draft phase II cohesive strategy documents contain key elements we recommended—such as laying out potential approaches for addressing the growing wildfire threat, estimating the costs associated with each approach, and identifying trade-offs. In the draft phase II document, the agencies reported that they expect to approve phase III of the cohesive strategy—entailing a national trade-off analysis—in late 2012 or early 2013. On the basis of our review of the phase I and draft phase II documents, it is not yet clear the extent to which phase III of the agencies' cohesive strategy will include the key elements we have recommended.

Other Lessons

Firefighters and officials from the Forest Service and other agencies offered additional observations about aspects of the Station Fire that offer potential lessons for wildland fire management. These observations include the following:

- *Defensible space.* Firefighters and agency officials told us that the Station Fire highlighted the importance of maintaining defensible space around structures in high-risk areas. Several officials and firefighters told us that, particularly in Big Tujunga Canyon, the lack of defensible space around certain homes made them difficult to protect. Forest Service officials acknowledged that at the time of the fire, residents whose homes were located closer than 100 feet to the boundary of the national forest or on land leased from the Forest Service were not allowed to create 100 feet of defensible space, the minimum required by California and Los Angeles County standards. Following the Station Fire, on December 17, 2009, the Forest Service's Pacific Southwest Region issued a memorandum directing forest supervisors in the region to increase the minimum requirement for defensible space around structures within and adjacent to national forests from 30 feet to 100 feet.[66] The Angeles National Forest is evaluating the effects of allowing private landowners to create up to 100 feet of defensible space on national forest land (up to 200 feet in some locations). The forest expects to decide whether to approve new defensible space standards by February 2012.
- *Protecting Mt. Wilson.* The Station Fire identified the importance of having an action plan to protect critical communications and other infrastructure on Mt. Wilson. According to some observers we spoke

to, the Forest Service and Los Angeles County did not have a previously agreed-upon plan of action to protect this area in the event of a fire and instead were debating what protective steps needed to be taken during the Station Fire. For example, some observers pointed to the fuel reduction work done during the Station Fire on Mt. Wilson as work that should have been done beforehand, given the location's importance. After the Station Fire, the Forest Service and Los Angeles County have recognized the importance of resolving this issue, and the Forest Service has taken actions to better prepare the Mt. Wilson area to withstand future wildland fires. In May 2010, for example, the Forest Service completed an environmental assessment to reduce fuels on 736 acres in the area by 2015. The Angeles National Forest has also decided to allow 300 feet of defensible space around the communications infrastructure and the Mt. Wilson Observatory, according to agency officials.

- *Communication with the public and media.* The Station Fire demonstrated the importance of timely and effective communication with the public and the media during such an event. According to some observers, the Forest Service did not adequately communicate important information to the public or the media about the fire's development or about the firefighting response, particularly after the extent of the damage became clear. Once questions about the adequacy of the Forest Service's response were raised, the Department of Agriculture and the Forest Service limited local agency officials' ability to communicate with the public and the media, which local officials said contributed to a perception that the Forest Service was covering up mistakes.

The Forest Service's Lessons Learned Document

On the basis of its experience with the Station Fire, as well as with its overall wildland fire management program, the Forest Service identified lessons from the Station Fire in an October 2010 *Station Fire Lessons Learned Report*. This report identified lessons applicable at local, regional, and national levels, including lessons regarding the role of night- flying aircraft and the need for improvements in the agency's ability to track aircraft water and retardant deliveries and to model fire behavior under certain conditions. In some cases, the agency identified actions that it planned to take to implement these lessons, including clarifying night-flying capabilities and aircraft use, improving procedures for ordering and dispatching firefighting assets, and allowing for increased defensible space on national forest lands in California. However, the report did not indicate whether actions were to be taken in implementing other lessons the agency identified and in many cases did not provide time frames for undertaking the actions discussed. For example, the document identified a number of areas where the Forest Service could standardize its procedures agencywide for ordering and mobilizing firefighting assets, but it did not describe whether or how the agency would do so. Forest Service officials told us that the document should not be considered a list of actions the agency plans to take and does not necessarily indicate that changes to policies or procedures are needed to address all the lessons identified. Nevertheless, we believe that the document presents an opportunity for the Forest Service to capitalize on the lessons offered by the Station Fire, and to provide information to Congress and the public about areas where the agency believes additional actions—such as changes to agency policies or procedures— are needed and where they are not. Such information could enhance Congress's and the public's understanding of how the agency views its own response to the fire and could improve their understanding of what changes to expect as a result.

CONCLUSIONS

The Station Fire was devastating in many ways—the tragic loss of two firefighters' lives, the destruction of dozens of homes, and the extensive damage to natural resources. Not surprisingly, given the significance of the event, views on the Forest Service's response to the fire vary widely. Some have questioned whether this devastation could have been prevented by a quicker or more robust Forest Service response to the fire, while others have complimented the agency, believing that its actions helped save thousands of homes in nearby communities. Answering such questions is a daunting challenge, both because so many firefighting decisions are made with imperfect information and under severe time constraints—and depend on firefighters' judgments about risk, safety, and effectiveness—and because an after-the-fact analysis such as ours offers only limited ability to determine whether different decisions should have been made given the information available at the time. Even less clear is whether, and to what extent, different decisions might have changed the outcome of the fire.

Nevertheless, important lessons can be learned from the Station Fire to help inform decisions about how to fight future wildland fires—although no steps can completely eliminate the threat of devastating fires. The Forest Service has taken actions to address some of the identified lessons learned from the Station Fire, such as updating its policies for using night- flying helicopters supplied by Southern California cities and counties (and undertaking a broader study of night-flying aircraft generally) and changing the Angeles National Forest's asset-ordering procedures to make the process more formal and systematic. Nevertheless, other issues identified in this report and in the Forest Service's own lessons- learned evaluation continue to warrant serious consideration and action. Specifically, the Forest Service has not yet clarified when its own firefighting assets should be ordered ahead of assets from other agencies. Similarly, the Forest Service can capitalize on other lessons learned from the Station Fire by taking actions that have the potential to improve its operations, clarify ambiguous operational processes, and address broader issues regarding its use of assets to fight fires, thereby laying the groundwork for improvements in its management of future fires.

Recommendations for Executive Action

To improve the Forest Service's response to wildland fires, we recommend that the Secretary of Agriculture direct the Chief of the Forest Service to take the following two actions:

- to clarify the Forest Service's intent and to reduce uncertainty about how its assets are to be used relative to those of other agencies, issue guidance describing when it expects its own firefighting assets to be used instead of contract or state and local agency assets, and
- document the steps it plans to take, and the associated time frames, to implement the lessons it identified in its review of the Station Fire.

Agency Comments and Our Evaluation

We provided a draft of this report for review and comment to the Department of Agriculture. In its written comments, the Forest Service, responding on behalf of the Department of Agriculture, generally agreed with our findings and recommendations. The Forest Service also provided technical comments, which we have incorporated as appropriate.

Anu K. Mittal
Director, Natural Resources and Environment

APPENDIX I. OBJECTIVES, SCOPE, AND METHODOLOGY

Our objectives were to (1) describe key events in the Station Fire and the Forest Service's response, including strategies, tactics, and assets used; (2) examine key issues arising from this response; and (3) identify lessons the Station Fire offers for wildland fire management in the future, including lessons specific to Southern California.

To describe the key events in the Station Fire and the Forest Service's response, we obtained and reviewed Station Fire records, including daily incident status summaries; daily plans of firefighting actions to be taken, known as incident action plans; and daily logs maintained by firefighters and decision makers. We also listened to recordings and reviewed transcripts and accounts of radio communications among firefighting officials and dispatchers. In addition, we obtained and reviewed documentation on orders placed for firefighting assets; this documentation was generated from the resource ordering and status system (ROSS), a computer software program that automates the resource ordering, status, and reporting process. We also reviewed agency and other reports that examined particular aspects of the fire, including those prepared by or in conjunction with the Department of Agriculture (the *Station Fire Initial Attack Review, November 2009* and the *Large Cost Fire Review for FY2009*), the Los Angeles County Fire Department (*Executive Review of Actions, Station Fire, November 2009*), other agencies that were directly involved in the Station Fire firefighting response (Orange County Fire Authority, *After Action Review: ORC Strike Team 1400C Structure Protection and Extreme Fire Behavior Event, Station Fire, Big Tujunga Canyon, Angeles National Forest, August 29, 2009*), and the Department of Agriculture's Inspector General. In addition, we reviewed Forest Service policies and guidance related to wildland fire incident management response, including the 2009 *Interagency Standards for Fire and Fire Aviation Operations*, wildland firefighting training manuals, interagency guidance on mobilizing firefighting assets to incidents, and the fire management plan for the Angeles National Forest. To obtain additional information, and to provide context for the documents reviewed, we interviewed numerous officials from the Forest Service and other agencies who were directly involved in the response. Specifically, we interviewed members of the incident management teams and other key support staff assigned to the Station Fire; senior officials from the nonfederal agencies in unified command with the Forest Service, including the Los Angeles County Fire Department, the Los Angeles County Sheriff's Department, and the Glendale, Los Angeles, and Pasadena Fire Departments; and agencies that were not part of the unified command but provided support through agency representation, including the California Department of Forestry and Fire Protection (CAL FIRE). We also interviewed Forest Service officials from agency headquarters, the Pacific Southwest Region, and the Angeles National Forest. In addition, we interviewed several residents who were affected by the fire, primarily residents of Big Tujunga Canyon. Rather than selecting individual residents to interview, we provided our contact information at a public meeting on the Station Fire, held in Altadena in April 2011, and interviewed those residents who subsequently contacted us. Some nonfederal officials and retired federal officials declined our request to interview them. While these officials may have provided additional perspectives or information on the fire, we believe that, given the number and variety of individuals we interviewed and the volume of documents we examined, our inability to interview these officials is unlikely to have substantially affected our findings, conclusions, or recommendations. In describing the Station Fire, we divided the fire into five phases to reflect changes in fire size and behavior, the risk to structures and resources, and the nature of the agency's response.

In addition, to gain a firsthand view of the Station Fire's location and the geographic features of the area, we visited several locations in the Angeles National Forest that were affected by the fire. These locations included the ignition site, Big Tujunga Canyon, Mt. Wilson, and others. During our visits we were accompanied by forest officials and firefighters involved in the response.

To examine the key issues raised by the Forest Service's response to the Station Fire, we reviewed agency and other reports, including those mentioned above, that evaluated aspects of the response and reviewed information provided at congressional meetings held in Southern California in October 2010 and April 2011. We also discussed key issues during our interviews with incident personnel; Forest Service officials; officials from nonfederal agencies; residents affected by the fire; and other interested parties, including retired Forest Service employees who did not participate directly in the incident. We also examined Station Fire records, including summaries, action plans, logs, recordings, and transcripts, as well as information from the ROSS database. Information from ROSS was particularly important for our assessment of the extent to which Forest Service assets were ordered rather than other, potentially closer assets. We obtained and analyzed asset-ordering information maintained in ROSS for two types of assets that were identified as slow to arrive at the incident—type 3 wildland fire engines and operations branch directors. For data about these assets, we requested the Forest Service to query ROSS for all orders placed for these assets from August 27 through August 30, 2009. We analyzed the data provided and determined the number of assets ordered and assigned by day and by ownership. However, limitations in the information available in ROSS on orders for firefighting assets precluded us from conclusively determining the extent to which agency assets were ordered ahead of assets available locally, because ROSS does not maintain information on assets that were available but not ordered. In addition, while orders in ROSS include the time and date by which the assets were needed on the fire, this information can be changed after the initial ROSS order is entered, precluding us from conclusively determining the extent to which assets arrived at or before the time for which they were initially requested. We assessed the reliability of the ROSS data used in our report by reviewing the methods of data entry into ROSS and determined that the data were sufficiently reliable for our use. We did not attempt to assess the compliance of individual firefighters or agency officials with applicable firefighting guidance, in part because responding to wildland fire requires a considerable degree of professional judgment.[67]

To identify lessons the Station Fire offers for future wildland fire management, we analyzed the information we obtained through Station Fire records and various interviews; we also reviewed several after-action reviews and narrative reports that identify suggestions for improvement, as well as Forest Service policy documents and previous GAO work on wildland fire management. In addition, during our interviews of federal and nonfederal agency officials and others, we specifically asked for participants' views on potential lessons offered by the fire. We also reviewed the Forest Service's *Station Fire Lessons Learned Report* prepared in October 2010, which describes lessons for the Forest Service at the national, regional, and forest levels, and in many cases identifies actions to be taken or actions that the agency has begun to take to address the lessons. We also reviewed a status update provided by Forest Service officials that identified both the changes that have been made as well as areas where changes are not expected.

We conducted this performance audit from December 2010 through December 2011 in accordance with generally accepted government auditing standards. Those standards require that we plan and perform the audit to obtain sufficient, appropriate evidence to provide a reasonable basis for our findings and conclusions based on our audit objectives. We believe that the evidence obtained provides a reasonable basis for our findings and conclusions based on our audit objectives.

APPENDIX II. THE STATION FIRE'S PROGRESSION AND FIREFIGHTING ASSETS USED OVER THE FIRE'S FIRST 9 DAYS, FROM AUGUST 26 THROUGH SEPTEMBER 3, 2009

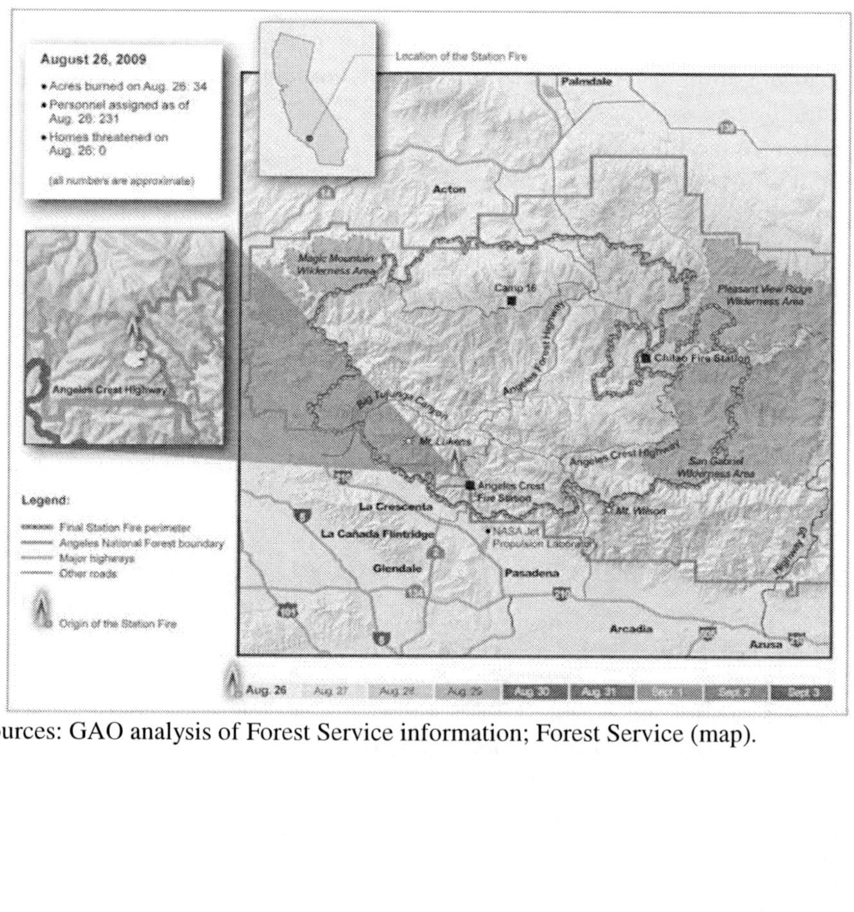

Sources: GAO analysis of Forest Service information; Forest Service (map).

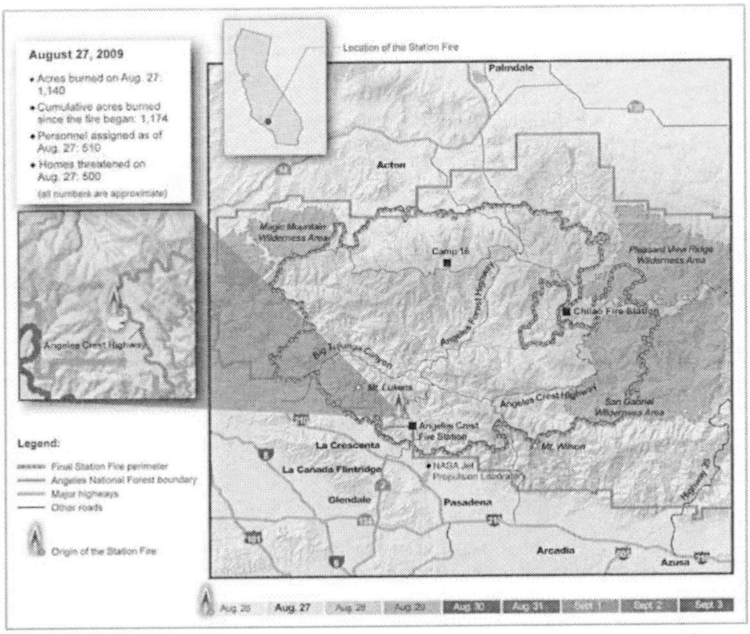

Sources: GAO analysis of Forest Service information; Forest Service (map).

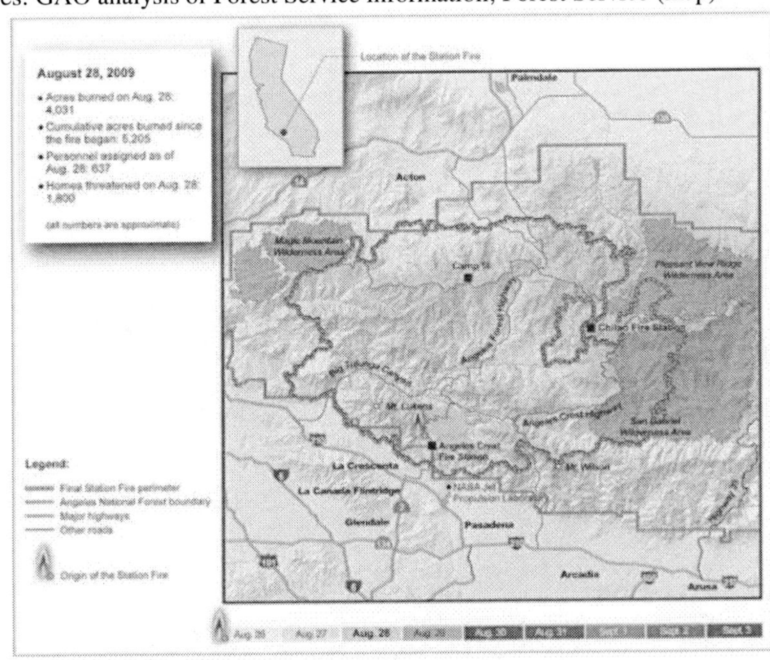

Sources: GAO analysis of Forest Service information; Forest Service (map).

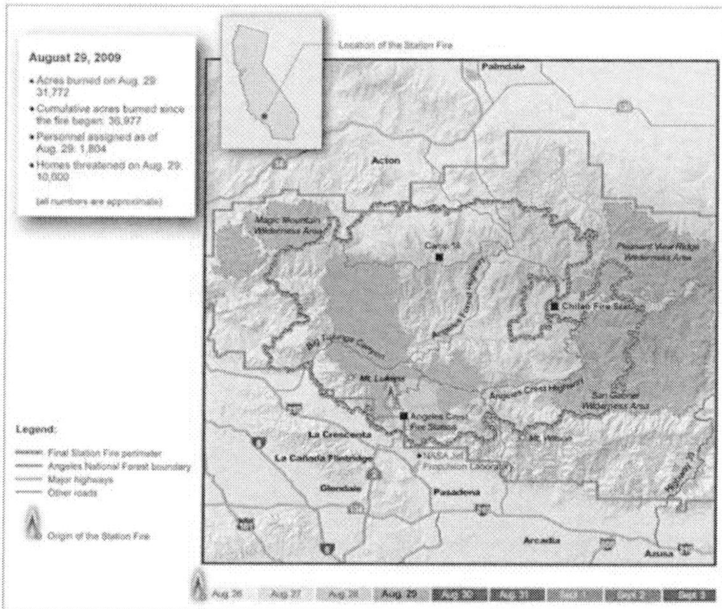

Sources: GAO analysis of Forest Service information; Forest Service (map).

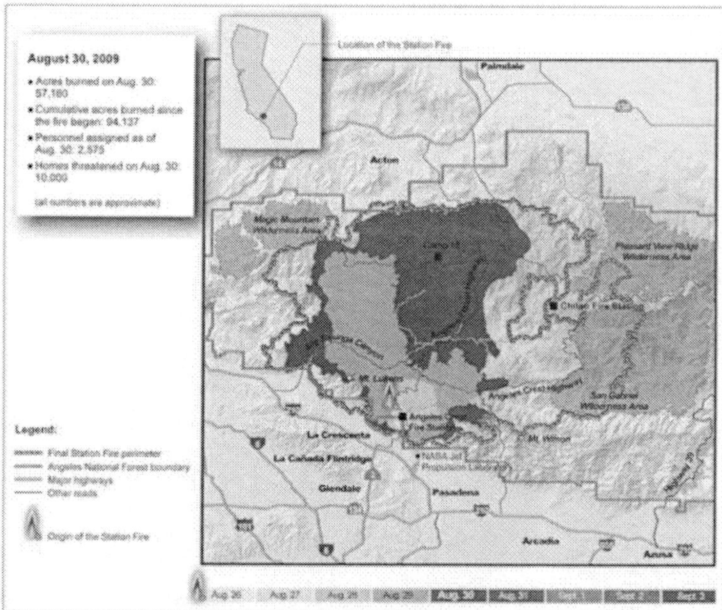

Sources: GAO analysis of Forest Service information; Forest Service (map).

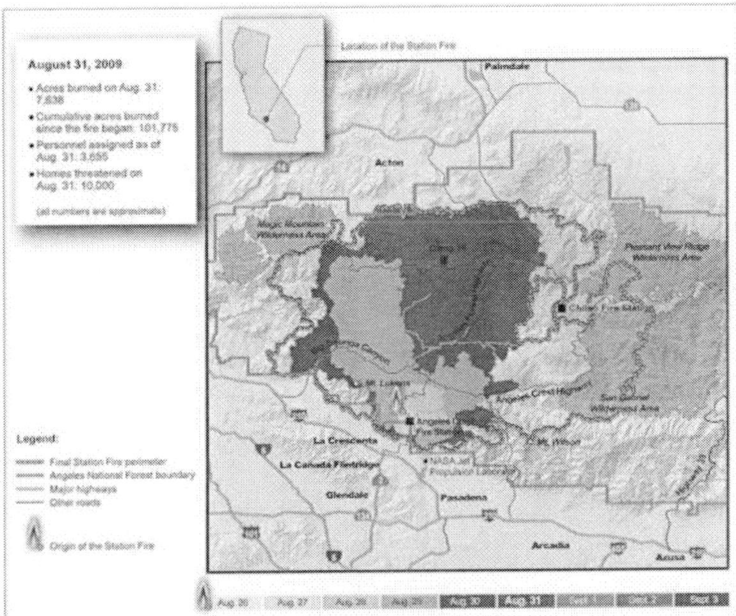

Sources: GAO analysis of Forest Service information; Forest Service (map).

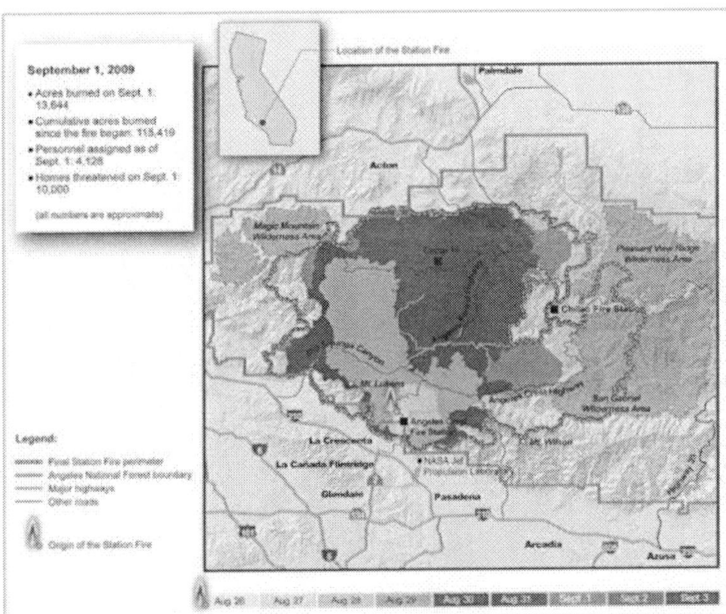

Sources: GAO analysis of Forest Service information; Forest Service (map).

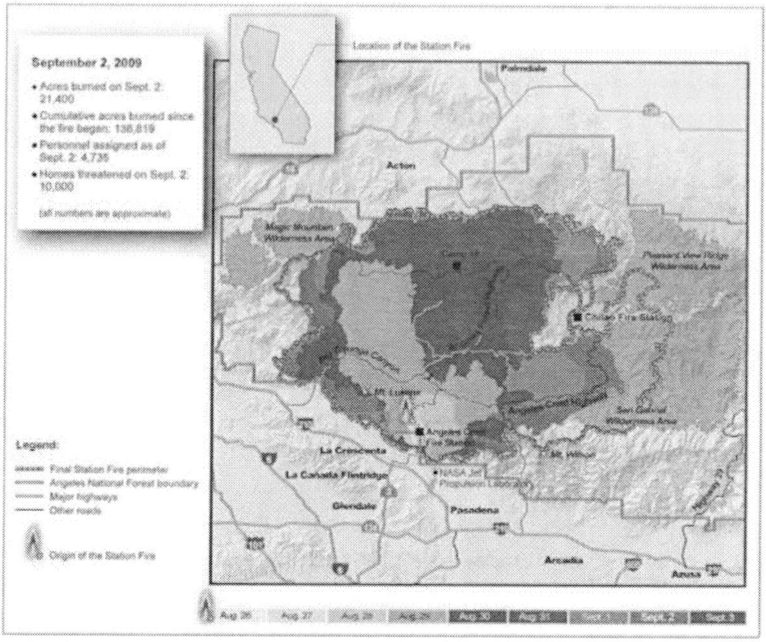

Sources: GAO analysis of Forest Service information; Forest Service (map).

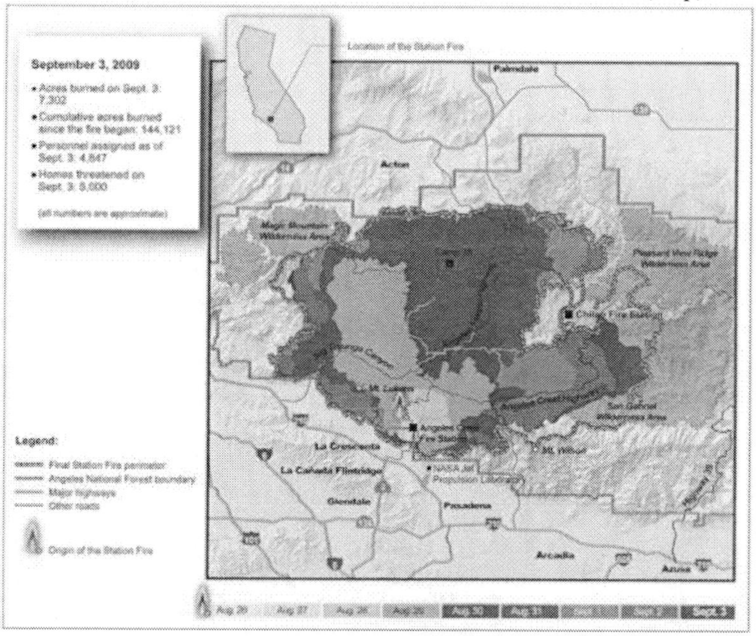

Sources: GAO analysis of Forest Service information; Forest Service (map).

End Notes

[1] Chaparral vegetation comprises several plant species, including chamise, scrub oak, and manzanita, whose thick, waxy leaves adapt them particularly well to drought.

[2] An operational period is the period of time scheduled for execution of a given set of tactical actions as specified in an incident action plan. Operational periods can be of various lengths, although usually not more than 24 hours.

[3] Eleven regional dispatch centers, called geographic area coordination centers, are located nationwide, each of which serves a specific geographic portion of the United States.

[4] The Forest Service obtains fixed-wing air tanker services through contracts with private industry.

[5] The fire was named the Station Fire because this location is near the Angeles Crest Fire Station in the Angeles National Forest.

[6] Flame length is a measure of fire intensity and refers to the distance from the tip of the flame to its base, which is generally at ground level. Thus, a fire that is burning at ground level and has a flame tip 6 feet high would have a flame length of 6 feet.

[7] Forest Service, *Angeles National Forest 2009 Fire Management Plan* (Arcadia, Calif.: June 2009).

[8] The specific objectives were to keep the fire below Lukens Ridge and above the Angeles Crest Highway.

[9] Los Angeles County helicopters are capable of flying missions at night, and, under certain conditions, these helicopters are allowed to fight fires at night on federal lands. The use of these helicopters in fighting the Station Fire is discussed later in this report.

[10] In Los Angeles, sunset was at 7:27 p.m. on August 26, 2009. According to Forest Service policy, helicopters that are not equipped to fly at night are required to return to their base no later than 30 minutes after sunset, which would have been 7:57 p.m. on that date. The other helicopters assigned to the Station Fire on August 26 were released at 6:10, 7:30, and 7:40 p.m. (three helicopters were released at 7:40). The two air tankers assigned to the Station Fire that evening returned to their base at 6:58 and 7:23 p.m.

[11] Firefighters told us they identified six spot fires but that there may have been others they could not see because of the terrain and the thick vegetation.

[12] The helicopters arrived at approximately 7:45, 9:14, 9:35, 10:00, 10:07, 10:15, and 10:27 a.m.

[13] Incident records confirm the use of air tankers and engines on these days but do not cite specific locations where they were used.

[14] Over the next several days, as the threat from the fire spread, additional jurisdictions entered into unified command with the incident management team. In addition to the Los Angeles County Fire Department, the following organizations entered into unified command during the incident: Glendale Fire Department, Los Angeles City Fire Department, Los Angeles County Sheriff's Department, and Pasadena Fire Department. Other agencies, including the California Department of Forestry and Fire Protection (CAL FIRE), also had representatives present and were involved in the response.

[15] The team assigned was California Interagency Incident Management Team 5.

[16] The DC-10 has a capacity of 12,000 gallons, and the 747 has a capacity of 24,000 gallons, substantially more than the approximately 3,000 gallons that a standard large air tanker can carry. The DC-10 made two drops this day, releasing more than 20,000 gallons.

[17] Incident records confirm the use of air tankers, crews, and bulldozers on these days but do not cite specific locations where they were used.

[18] This was one of several structure protection groups created by the incident management team.

[19] Defensible space is the area around a structure, generally between 30 and 100 feet wide, in which the removal of flammable materials and reduction of vegetation are undertaken to reduce the likelihood that a wildland fire will damage or destroy the structure. For more information on defensible space, see GAO, *Technology Assessment: Protecting Structures*

and Improving Communications during Wildland Fires, GAO-05-380 (Washington, D.C.: Apr. 26, 2005). California law (Public Resource Code § 4291) requires homeowners to maintain 100 feet of defensible space around a home, although generally not beyond the property line.

[20] In a temperature inversion, cooler air higher in the atmosphere acts as a cap, preventing warmer air near the earth's surface from rising.

[21] Incident records we reviewed do not indicate the precise time of the evacuation. According to firefighters, the process for initiating an evacuation was communicated to the appropriate officials immediately after the fire reached the trigger point. Dispatch communications recorded at approximately 11:15 a.m. show that the evacuation had started by that time.

[22] Orange County Fire Authority, *After Action Review: ORC Strike Team 1400C Structure Protection and Extreme Fire Behavior Event, Station Fire, Big Tujunga Canyon, Angeles National Forest, August 29, 2009* (Orange County, Calif.; Oct. 1, 2009).

[23] The Los Angeles County Fire Department operates nine fire camps located throughout the county, including within the Angeles National Forest, to assist in wildland fire suppression, reduce vegetation, and assist with minor building construction.

[24] Los Angeles County Fire Department, *Executive Review of Actions: Station Fire* (Los Angeles: Nov. 17, 2009).

[25] The Martin Mars carries water or suppression gel that is dropped directly on the fire, rather than fire retardant that is dropped ahead of the fire.

[26] A wildland fire is considered contained when a fuel break around the fire has been completed. This break may include natural barriers or manually or mechanically constructed line.

[27] The Martin Mars was accompanied by a helicopter to lead it over the fire, along with an official assigned by the Forest Service who was in charge of directing the tanker and coordinating with other incident officials.

[28] The aircraft was required to land at its base by 30 minutes after sunset, according to the pilot. On August 26, 2009, official sunset at the tanker's base at Lake Elsinore was at 7:22 p.m., meaning that the plane had to land by 7:52 p.m. The official directing the Martin Mars estimated that it would have taken about 18 minutes for the aircraft to fly from the Morris Fire to the Station Fire, identify a drop location, and make the drop; and about 33 minutes for it to then return to its base at Lake Elsinore.

[29] A hand crew typically comprises 18 to 20 firefighters.

[30] Recordings of the conversation between the incident commander and the dispatcher show that when the incident commander was asked what time he wanted the air tankers to arrive, he said, "I know it's not possible but, you know, 0700."

[31] The three air tankers were entered into ROSS as request numbers A-16, A-17, and A-18.

[32] The geographic area coordination center for Southern California, located in Riverside, provides interagency logistical support for wildland fire incidents, including processing requests for firefighting assets.

[33] Department of Agriculture, Office of Inspector General, *Report of Investigation: Unidentified Employees, Los Angeles, CA (Los Angeles County),* File No. SF-0801 -0638 (San Francisco: June 2011).

[34] Angeles National Forest dispatch personnel told us they have the authority to divert air tankers from one fire to another within the same forest.

[35] The three air tankers requested around 12:30 a.m. were reentered into ROSS as request numbers A-35, A-36, and A-37; the air tanker requested at 8:21 a.m. was entered into ROSS at the same time as the others—approximately 8:40—as A-38.

[36] Our analysis did not include the Martin Mars air tanker. Unlike most air tankers, the Martin Mars does not drop retardant (it drops water or suppression gel); it therefore would be used in different circumstances than air tankers dropping retardant. According to an Angeles National Forest official, for the Martin Mars to have been mobilized to the fire, the incident commander would have had to specifically request it. Because the incident commander did not specifically request the Martin Mars, we did not analyze the aircraft's availability.

[37] The crews for two of these five air tankers went off duty at 7:59 p.m., the crew for the third at 8:00, and the crew for the fourth at 8:18. The Forest Service did not provide a record of when the crew for the fifth air tanker, which was located in Fresno that night, went off duty on August 26, but the tanker base manager said that the air tanker was assigned to the Big Meadow Fire on the morning of August 27.

[38] The Forest Service has stated that it does not require its contractors to provide more than one crew for the air tankers. Although doing so would allow air tankers to operate from sunrise to sunset, the agency has stated that extending the hours the aircraft operate would not be safe because of the age and maintenance requirements of the aircraft.

[39] The 10-hour uninterrupted-rest restriction also applied to the crews of the air tankers located in Fresno and San Bernardino.

[40] The base manager said there was only one fueling truck available that morning, which is typical at the base unless a large fire is burning nearby. Because the Station Fire had just started the day before, base officials had not yet ordered an additional fueling truck.

[41] CAL FIRE flight crew rest requirements provide that within any 24 hour period, pilots shall have a minimum of 10 consecutive hours off duty.

[42] According to a National Wildfire Coordinating Group handbook, the air tactical group supervisor is responsible for making tactical recommendations to appropriate incident personnel, but it is the responsibility of the operations section chief to determine asset needs and request additional assets. If a fire does not have an operations section chief— for example, if the fire is being managed by a type 3 incident management team, as the Station Fire was on August 27—responsibility for final approval for requesting assets lies with the incident commander, according to a senior Forest Service official.

[43] A trainee may be assigned to work with a person qualified to perform a specific role, such as incident commander, in order to receive additional experience before being certified to perform that role on future incidents. A trainee meets all the required training and experience for the position and performs such training assignments in order to complete his or her qualifications for the position.

[44] In a review of costly fires that occurred in 2009, a panel convened by the Department of Agriculture reported that very large air tankers were ordered on the Station Fire in part because of political pressure from outside the agency and that Angeles National Forest officials said they would not have ordered these aircraft had they not been pressured to do so. See Secretary of Agriculture's Independent Large Cost Fire Review Panel, *Large Fire Cost Review for FY2009* (Washington, D.C.: August 2010). Angeles National Forest officials we interviewed, including one who was involved in approving the use of very large air tankers during the Station Fire, said they were not interviewed by the review panel and that they did not agree with this statement.

[45] The Forest Service contracted with the National Aeronautics and Space Administration to evaluate the safety and utility of the DC-10 and Boeing 747 for firefighting operations. In March 2009, that agency recommended the aircraft not be used in "steep or rugged terrain" unless the drop could be made with minimal maneuvering, and a lead plane and adequate ground clearance are available.

[46] See appendix I for the details of our scope and methodology, including a discussion of these limitations.

[47] Engines are rated according to their size and mobility and are classified according to type. Among engines typically used to fight wildland fires, capacity for pumping water or foam increases from type 7 engines, which have the least capacity, to type 3 engines which have the greatest. Type 1 and 2 engines are typically used to fight structural fires and are not commonly used in wildland firefighting.

[48] A fire planning document created by the Angeles National Forest identified many locations across the forest where bulldozer or hand lines could be established in case of a fire. The document showed that a bulldozer line could be established in Big Tujunga Canyon where

the line was started during the Station Fire but that bulldozers would likely not be able to reach the canyon bottom and that firefighters would have to complete the line by hand.

[49] One strike team was assigned to the eastern part of the canyon, near Big Tujunga Dam; one strike team was assigned to the Vogel Flats area, in the middle of the canyon; and one strike team was assigned to the western portion of the canyon near Delta Flats. Vogel Flats and Delta Flats are areas of the canyon with a higher concentration of residences.

[50] Removing vegetation and other flammable materials to create defensible space around a structure is the responsibility of the homeowner, but firefighters may take such steps if they have time and it is safe to do so.

[51] Incident documents for August 29 and August 30 indicate that 3 and 18 homes, respectively, were destroyed.

[52] Incident management teams determine when an area should be evacuated and request that an evacuation notice be issued; the responsible local law enforcement agency issues evacuation notices and carries out the evacuation.

[53] Although we reviewed key events related to the evacuation of Big Tujunga Canyon, we did not assess whether the manner in which evacuations took place followed applicable state and local law and policy.

[54] The Forest Service uses fire behavior models to help predict the intensity of a fire and how quickly it will spread, but the ability of the models to predict plume-dominated fire behavior is limited, according to a Forest Service document.

[55] The Department of Agriculture's Office of Inspector General examined the Forest Service's plans for recruiting, training, developing, and retaining fire management personnel, reporting in March 2010 that the agency lacked a workforce plan specific to firefighters, despite the relatively high number of eligible retirees among those in positions critical to firefighting and the agency's own expectations of an increase in the size and number of fires it will be responsible for suppressing. The Forest Service has implemented 4 of the Inspector General's 20 recommendations and is taking steps to implement the others, but these steps have not been completed. See Department of Agriculture, Office of Inspector General, ***Forest Service's Firefighting Succession Planning Process***, Audit Report 08601-54-SF (Washington, D.C.: Mar. 31, 2010).

[56] Forest Service, *Station Fire Lessons Learned Report* (Washington, D.C.: October 2010). Of the five lessons described here, the Forest Service's lessons learned document discusses four: those relating to night-flying aircraft, firefighting asset ordering and dispatching systems, tracking of aircraft water and retardant deliveries, and fire behavior modeling. Other lessons included in the Forest Service's lessons learned document relate to wildland-urban interface communities, defensible space around structures, interagency agreements, incident documentation and record keeping, after action reviews, communication with the media and the public, incident management team transitions, firefighter training, and interagency coordination. (This document is available at http://wildfirelessons.net/documents/Station_Fire_Lessons.pdf.)

[57] The other entities are Kern County, Orange County, Santa Barbara County, Ventura County, and the city of San Diego.

[58] The previous Forest Service policy limited the use of night-flying helicopters to multiengine aircraft.

[59] *2011 Annual Operating Plan for Cooperative Fire Protection Agreement between Consolidated Fire Protection District of Los Angeles County and U.S. Forest Service Angeles National Forest.*

[60] *2009 Annual Operating Plan for Cooperative Fire Protection Agreement between Angeles National Forest and Consolidated Fire Protection District Los Angeles County Fire District.*

[61] Forest Service and Fire Program Solutions, LLC, *Helicopter Night Operations Study* (San Dimas, Calif., and Sandy, Ore.: 2010).

[62] GAO, *Forest Service: Continued Work Needed to Address Persistent Management Challenges*, GAO-11-423T (Washington, D.C.: Mar. 1 0, 201 1). For additional information on FPA, see GAO, *Wildland Fire Management: Interagency Budget Tool Needs Further Development to Fully Meet Key Objectives*, GAO-09-68 (Washington, D.C.: Nov. 24, 2008).

[63] GAO has issued numerous reports and recommended more than 50 actions the Forest Service and Interior agencies could take to improve wildland fire management. For more information on the agencies' efforts over the past decade in this regard, see GAO, *Wildland Fire Management: Federal Agencies Have Taken Important Steps Forward, but Additional, Strategic Action Is Needed to Capitalize on Those Steps*, GAO-09-877 (Washington, D.C.: Sept. 9, 2009).

[64] Pub. L. No. 111-88 § 503, 123 Stat. 2971 (2009).

[65] Department of the Interior and Department of Agriculture, *A National Cohesive Wildland Fire Management Strategy* (Washington, D.C.: March 2011).

[66] According to the Forest Service, forest supervisors have the authority to authorize greater clearance if warranted by conditions at a specific location.

[67] According to *Interagency Standards for Fire and Fire Aviation Operations*, "fire operations doctrine does not consist of procedures to be applied to specific situations so much as it sets forth general guidance that requires judgment in application."

In: The Station Fire ISBN: 978-1-62081-084-2
Editors: E. L. Donahue and D. M. Elmore© 2012 Nova Science Publishers, Inc.

Chapter 2

FIRE AND AVIATION MANAGEMENT STATION: FIRE INITIAL ATTACK REVIEW*

United States Department of Agriculture

EXECUTIVE SUMMARY

On August 26, 2009, at 3:24 p.m., the Angeles National Forest (USDA Forest Service, Pacific Southwest Region) received a call for initial attack response to a fire that had ignited on the forest near the Angeles Crest Fire Station, on the Angeles Crest Highway. The terrain below the Angeles Crest Highway is rugged and steep. The slope percentage from the highway to the creek bottom ranges from 33 percent to 67 percent along the half mile section of the highway centered on the fire station. There are several narrow ravines facing east along this section of highway. The slopes below the highway range from 900 to 1,680 feet in length. The vegetation was mature mixed chaparral, standing six to eight feet tall, at least 50 years old, and extremely flammable. Ultimately, the fire grew in size to over 160,000 acres, leading to the loss of forest resources, private property, and human life. At the direction of the Chief of the USDA Forest Service, a review was undertaken for the period from initiation of initial attack on August 26 to August 28, 2009, at 2:00 p.m., when the fire was placed under the command of a Type I Incident Management Team.

* This is an edited, reformatted and augmented version of the United States Department of Agriculture Publication, dated November 13, 2009.

A five-person review panel, guided by incident documentation and interviews conducted with initial attack incident commanders and Angeles Forest personnel, performed a review that reconstructed the initial attack phase of the Station Fire[1] from a foresight perspective. The review process focused on conditions at the time of the incident including terrain, weather, vegetation conditions, potential fire behavior, suppression resource effectiveness, and exposure of firefighters to the hazards of wildland fire. From this foresight perspective, decisions were reconstructed in terms of how they appeared to incident commanders at the time, given the information available and the uncertainties present in the management situation.

The primary objectives of the review were to:

- Determine if the Angeles National Forest preparedness actions before the Station Fire were in alignment with existing plans (staffing/ action guides, land and resource management plan, fire management plans) and cooperative agreements;
- Determine if the decision process and actions taken during initial attack including key decisions during the initial attack were reasonable and prudent in view of the known environment, current information, and other circumstances existing at the time of the event; and
- Examine the role of the agency administrator in providing leader's intent and input to strategic decisions for the incident initial attack.

Terrain below the Angeles Crest Highway is extremely rugged and steep.

The review panel found that:

- The Angeles National Forest had in place at the time of the incident an up-to-date staffing and action guide for initial attack.
- The actions taken by the Angeles National Forest and the Forest Supervisor with respect to overall incident objectives—controlling the fire at the smallest acreage practicable consistent with firefighter safety considerations, were consistent with the forest's land management plans.
- The origin of the Station Fire was in extremely rugged terrain with limited opportunities for safe suppression activities by ground-based suppression resources.
- The dry, dense brush in the area of the fire was at high risk for potentially extreme fire activity and at a level that posed unacceptable risk to firefighters.
- Firefighters made cooperative efforts to engage the fire at critical points during the daylight phase of initial attack. Control of the incident was prevented because of a spot fire that occurred in an inaccessible location with limited visibility and thick, tall brush.
- The ordering and assignment of firefighting resources to initial attack was appropriate and consistent with accepted fire management practices. Additional ground tactical resources would not have improved the effectiveness of operations because they could not be safely deployed.
- Incident management decisions made during the review period were consistent with generally accepted incident management practices. Decisions made by initial attack incident commanders reflected sound judgment of the operational situation and were prudent with respect to firefighter effectiveness, safety, and suppression resource deployment.
- The review panel found no evidence or indication that initial attack incident commanders felt unduly constrained to inappropriately reduce direct suppression costs.
- In conclusion:
  - Incident managers during the initial attack phase of the Station Fire acted in a manner consistent with best professional practices as accepted by wildland firefighting agencies, and
  - Deployed suppression resources under conditions where firefighters would be safe and effective.

Active fire below the Angeles Crest Highway coupled with the topography make accessibility by ground crews impossible.

In light of the extremely challenging topography encountered during initial attack and the highly volatile fire and vegetation conditions, incident commanders were reasonable and prudent in not exposing firefighters to actions that would have been ineffective and compromised their safety.

INTRODUCTION

The safe, effective, and successful initial attack of wildland fire is vital. As an agency, the USDA Forest Service is successful in its initial attack of wildland fires nearly 98 percent of the time. The Angeles National Forest Land and Resource Management Plan mandates the forest take aggressive action on all unplanned ignitions. During the 2008-2009 fire season, the forest experienced only two escapes of the nearly 200 fires that occurred. The 160,500 acre Station Fire was one of the two fires to escape initial attack.

Initial attack is one of the most difficult, hazardous phases of fire management. Improving initial attack practices means not only improving effectiveness of suppression but also improving the management of risk to those who intervene as part of fire operations and are thereby exposed to the inherent risks of wildland fire. Each year, wildland firefighters die in the line of duty – approximately 20 annually. Incident managers have a responsibility to protect the public as well as firefighters from exposure to wildland fire that could prove hazardous and even fatal.

Angeles National Forest has long history of firefighter fatalities, especially those that occurred when fire came from below the firefighters.

The Angeles National Forest has a long history of firefighter fatalities, especially those that occurred when fire came from below the firefighters, overrunning them, as was the case during the Canyon Fire (1968), the Loop Fire (1966), and the Glen Allen Fire (1993). Hot, dry weather, with very steep topography, and the vegetation conditions prone to this particular forest made firefighters on the Angeles National Forest acutely aware of the dangers of approaching a fire from above.

Scope and Focus of Review

This review is entirely focused on the period of time between discovery and report of the Station Fire on the afternoon of August 26, 2009, at 3:24

p.m., to transfer of command of the incident to a Type I Incident Management Team on August 28, 2009, at 2:00 p.m. The review addresses neither the performance of the Type I Incident Management Team nor the predisposing vegetative and societal factors that contributed to the Station Fire. This review does not address the performance of the non-federal assistors and cooperators who were also engaged in this incident.

The primary objectives of the review were to:

- Determine if the Angeles National Forest preparedness actions before the Station Fire were in alignment with existing plans and cooperative agreements;
- Determine if the decision processes and actions taken during initial attack, including key decisions during the initial attack, were reasonable and prudent in view of the known environment, current information, and other circumstances existing at the time of the event; and
- Examine the role of the agency administrator in providing leader's intent and input to strategic decisions for the incident initial attack.

Review Panel Composition

Review Panel was composed of the following individuals:

- James E. Hubbard, Team Leader, Deputy Chief, State and Private Forestry, USDA Forest Service
- Tom Harbour, Director, Fire and Aviation Management, USDA Forest Service
- John B. Tripp, Chief Deputy, Emergency Operations, Los Angeles County Fire Department
- Ken Pimlott, Deputy Director, Chief of Fire Protection, California Department of Forestry and Fire Protection (CAL FIRE)
- Donald G. MacGregor, Ph.D., Senior Scientist, MacGregor and Bates, Incorporated

Fire above and below Angeles Crest Highway on August 27 after spot fire occurred.

From discovery of the fire to approximately 1:10 p.m., on August 27, 2009, the Angeles National Forest personnel were entirely responsible for the management of the fire. During this period, the Station Fire was not under unified command, and suppression resources supplied by the California Department of Forestry and Fire Protection (CAL FIRE) and the Los Angeles County Fire Department (County) were provided solely in a role of assistance. Unified command was established with Los Angeles County Fire Department at approximately 1:10 p.m. on August 27, 2009. During this review, the role of the two review panel members from the Los Angeles County Fire Department and CAL FIRE organizations was to provide independent technical or subject matter expertise and to analyze the wildland firefighting strategies and tactics used during initial attack, from an outside perspective. Their input was based on the information provided by the Angeles National Forest and through interviews conducted with initial attack incident commanders and other forest personnel. Neither of these team members was in a role to evaluate conformance of initial attack actions with USDA Forest Service policies, directives, and/or guidelines.

Review Process

The general approach to this review is adapted from the "After Action Review" format described in the National Wildland Fire Coordinating Group (NWCG) Incident Response Pocket Guide (PMS #461).

Individuals Interviewed

Interviewee	Unit	Role on Incident
INT #1	Angeles NF	Fire Officer
INT #2	Angeles NF	Incident Commander (Type III)
INT #3	Angeles NF	Incident Commander (Type IV)
INT #4	Angeles NF	Incident Commander (Type III)
INT #5	Angeles NF	Incident Commander Trainee
INT #6	Angeles NF	Fire Investigator
INT #7	Angeles NF	Division Supervisor

In addition, the review takes advantage of a risk science-based approach that recognizes the complexity and uncertainty in hazardous situations, and the need for incident managers to assess the likely effectiveness of suppression resources in a given situation, as well as to assess the risks to incident responders.[2, 3]

Wildland fire suppression is a hazardous undertaking, as such, wildland fire management decision making is always done under conditions of uncertainty. This means that a review of fire management decision making cannot take, as its reference or starting point, the outcome of an incident. To do so is misleading and lends to a hindsight-bias by which the outcomes can appear inevitable from preceding conditions and decisions. To avoid a hindsight bias, this review was focused on reconstructing the decision situation as it appeared to key decision makers at the time, during initial attack. A fore-sight perspective accounts for the uncertainty that is inherent in wildland fire incident command as well as the risks to which firefighters are exposed. From this perspective, the review is then appropriately focused on:

- Whether the decisions made on the incident were reasonable given the situational conditions, uncertainties present, and the information available at the time, and
- Whether similar individuals, with similar professional backgrounds, and experience, operating under similar circumstances would come to the same general conclusions about the best approach for dealing with the fire.

A key consideration given in the review process was to the exposure of firefighters to the inherent risk of wildland fire, the appropriateness of exposure given the environmental conditions (e.g., weather, terrain, vegetation, etc.), the fire behavior, consideration to the values imminently threatened and at risk, and ultimately the firefighters' ability to be effective in suppressing the fire. Fire management decisions are risk based and are a tradeoff between the damage a fire might do and the harm it might inflict to firefighters who intervene to mitigate those risks. For every firefighter assigned to take action on a fire incident, risk is transferred from the resource threatened to the firefighter. This review looked at if and when that transfer of risk was reasonable and appropriate during the initial attack phase of the Station Fire; and when it was not, did incident managers act prudently in making decisions that reduced firefighters' exposure to harm.

The review was initiated by collecting information and documentation directly pertaining to the incident including:

- Dispatch Logs or Run Cards
- Weather Data (predicted, observed, national fire danger ratings)
- Pre-attack Plans
- Decision Documentation (ICS-214 and Logs)
- Maps
- Briefing Packages
- Situation Reports (ICS-209)

This information was used as a guideline for interviewing incident managers in key decision- making roles during the review period. From those interviews, as well as appropriate incident documentation, incident decisions were reconstructed in terms of:

a. Environmental conditions that pertain to the fire and fire context (e.g., fire behavior, weather, vegetation), and
b. Management events that represent responses to the fire.

The incident reconstruction was used to develop the findings and recommendations presented in this report.

INCIDENT RECONSTRUCTION

The incident reconstruction was based on a number of information sources including incident documentation, interviews with initial attack incident commanders, visits by members of the review panel to the site of the initial attack, and subsequent information requests made to the Angeles National Forest management and staff. In addition, the review panel requested a review of the conditions present at the site of initial attack relevant to factors that influenced the fire behavior associated with the spot fires below the Highway. This evaluation was completed by Dr. David Weise of the USDA Forest Service, Pacific Southwest Research Station in Riverside, California. (See Appendix A)

Highlights of the report are as follows:

General Setting and Terrain Conditions

The ignition occurred along the Angeles Crest Highway (Highway 2) at approximately milepost 29, near the Angeles Crest Fire Station (Lat 34° 15' 3.77"N; Long 118° 11' 41.51"W). The fire investigation placed the exact point of ignition adjacent to a roadside turnout and indicated it was human caused. The terrain above and below the Angeles Crest Highway at this location is rugged and steep. The slope percentage from the Highway to the creek bottom ranges from 33 percent to 67 percent. Appendix B shows photographs of the site of initial attack taken from several vantage points, including a point on Civilian Conservation Corps (CCC) Ridge looking toward the initial attack area. As the photographs indicate, topographical conditions in this area are extremely difficult, if not impossible, to traverse without a high degree of exposure to hazard. In addition to the hazards of rolling rocks and unstable footing, firefighters attempting to access points below the Highway would be in an upslope position with respect to fire below them and in brush conditions with extremely limited or no visibility. In these conditions, safety zones and escape routes could not be established.

Along Angeles Crest Highway at Turnout.

Firefighters attempting to access points below the Highway would be in an upslope position with respect to fire below them—safety zones and escape routes could not be established.

Weather Conditions

Weather conditions at the time of Initial attack and into the evening of August 26 tended to be warm and dry with relatively light winds. Weather observations at the Clear Creek weather station (approximately 3.2 miles away) indicated a temperature of 89 degrees Fahrenheit, a relative humidity of 11 percent, and wind speed of 10 mph. The spot weather forecast produced at 5:20 p.m., in the late afternoon and early evening, predicted night time temperatures around 74 degrees Fahrenheit, and a maximum relative humidity of 17 percent. Maximum observed relative humidity at the two weather stations in the area on the night of August 26 did not exceed 17 percent. At 8:40 a.m. on August 27, temperatures were 89 degrees Fahrenheit with relative humidity of 9 percent.

Vegetation Conditions

Vegetation conditions are a critical element in determining the hazardousness of exposing firefighters in incident operations. The observed afternoon vegetation moisture content was typical for this time of year. However, the fact that the relative humidity was not predicted to increase very much on the night of August 26 meant that the vegetation moisture would stay relatively low throughout the night and that fires would continue to burn actively through the night. Live vegetation moisture content was normal for this time of year based on long-term averages. However, the moisture content in chamise was approaching 60 percent, which is considered a critical level for fire behavior. Also of importance was the percentage of dead versus live vegetation. Dead vegetation is particularly volatile and can lead to extreme fire behavior. Percentage of dead vegetation in the area of initial attack ranged from 50 percent to 70 percent.

Potential Fire Behavior

Initial Attack on the Station Fire occurred in a context of highly volatile vegetation conditions with potentially extreme fire behavior. Guidelines for suppression activities that are appropriate for different levels of fire behavior have been developed over the years and are a basis for determining the risk associated with suppression operations. Flame length, rate of spread, and fire intensity are the fire behavior indicators typically used. In the case of the initial attack on the Station Fire, vegetation conditions were such that flame lengths in excess of 15 feet could be expected, accompanied by very high rates of spread. Under these conditions, fire behavior is extreme and not amenable to attack by hand crews or fire engines.

Vegetation conditions are a critical element in determining the hazardousness of exposing firefighters to incident operations.

INCIDENT SYNOPSIS

The first initial attack dispatch on the Station Fire occurred at approximately 3:24 p.m., on August 26, 2009. Simultaneously Los Angeles County Fire Department and Angeles National Forest personnel responded,[4] with additional air resources provided by the nearby Morris Fire. On arrival, incident responders found the fire burning on the upslope side of the Angeles Crest Highway, adjacent to a roadside turnout, in hazardous conditions (see Appendix C for pictorial representation of IA site). The terrain at the site of initial attack was extremely rugged and steep. Firefighters engaged the fire with handcrews, engines, and aircraft. Operational decisions were based on where resources could be both safely and effectively positioned. Hose lines, aircraft, and handcrews were deployed on the fire. Ground-based resources (crews and engines), accompanied by air support, were able to make progress on the upslope portion of the fire throughout the afternoon. Only a portion of the resources on the incident could be used, however, due to limited opportunities for deployment posed by the hazardous terrain, brush conditions with extremely limited or no visibility, and potential fire behavior conditions.

These undeployable resources were released once it was obvious to the incident commander that they could not be utilized safely and effectively.

Spot fires occurred on the down slope side of the Highway within minutes after responders arrived. Temperatures at this time of the day were reaching 100 degrees, and the relative humidity was less than 10 percent, making the probability of ignition on the down slope aspect almost certain and the potential fire behavior extreme. Engines, handcrews, and/or aircraft were able to attack some spot fires; however, there was at least one known spot fire which was inaccessible for attack. Air resources were unable to extinguish fires alone and would have required ground resources to assure effectiveness.

The upslope portion of the fire continued to respond to control efforts using the appropriate resources. Late in the evening of August 26, flying embers produced by the up slope fire ignited at least one significant spot fire below the Highway in an area inaccessible by ground resources. The on-scene assets, including engine crews, hand crews, and leadership, assessed the situation regarding the spot fires and determined that without appropriate anchoring and mitigation, approaching the fire from above was neither safe nor feasible. By 2:10 a.m. on the morning of August 27, 2009, several spot fires had occurred on the down slope side of the Highway. Due to the increased fire activity, orders were placed for additional air resources and ground resources with a reporting time of sunrise. Effective work continued on the upslope portion of the fire and assessment continued for the fires down slope of the highway. By 8:30 a.m. on the morning of August 27, the fire was exhibiting extreme fire behavior with fire burning both upslope and down slope of the Highway. At 11:00 a.m., firefighters were ordered off the line due to extreme fire behavior and the unsafe conditions. At approximately 1:10 p.m., the Angeles National Forest and Los Angeles County Fire Department entered into unified command. At 1:32 p.m. on the afternoon of the August 27, a Type I Incident Management Team was ordered. The Type I Incident Management Team assumed command of the fire on August 28, 2009, at 2:00 p.m.

Initial Attack on the Station Fire occurred in conditions that were extremely hazardous and in which there were limited opportunities to engage the fire operationally. The spot fires that occurred down slope of the Highway on the evening of August 26 required ground resources to assure suppression because air resources alone were insufficient to achieve control. The spot fire situation continued to worsen throughout the evening of August 26 and early morning of August 27. Sufficient fire resources were available to incident managers; however, the inaccessibility of the fire and excessive hazard created by the terrain and dry, dense brush conditions limited opportunities to affect fire spread down slope of the Highway. Conditions were not safe for ground crews to approach the fire from above.

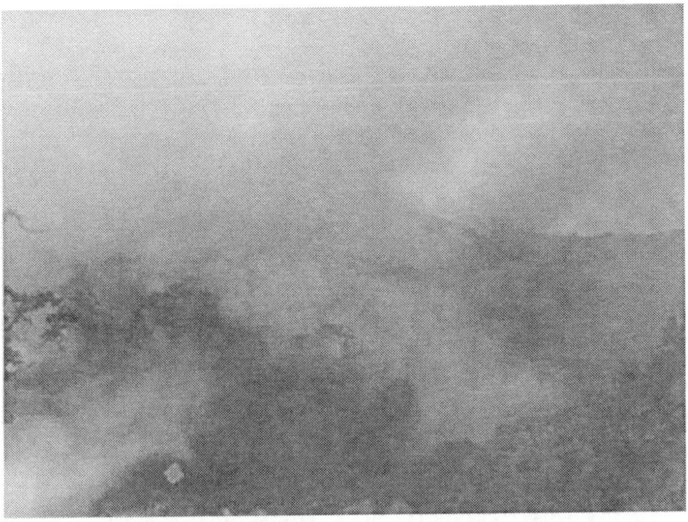

Operational decisions were based on where resources could be both safely and effectively deployed.

Map showing location of spot fire and inaccessible, rugged terrain below the Highway.

Sufficient resources were available to incident managers; however, the inaccessibility of the fire and excessive hazards limited opportunity to affect fire spread.

FINDINGS OF THE REVIEW PANEL

The review panel was asked to determine whether the decision processes and actions taken during initial attack were reasonable and prudent in view of the known circumstances existing at the time of the event. Based on information obtained from incident documentation as well as interviews with incident managers, the review panel identified several key findings and found decisions were reasonable and prudent based on the following findings:

Preparedness Actions of the Angeles National Forest

At the time of the incident, the Angeles National Forest had in place an up-to-date staffing and action guide for initial attack. In addition, the forest had established preparedness plans with respect to readiness, communications, and interagency cooperation. The forest had coordinated actions in this area with the Los Angeles County Fire Department, who assisted in the initial attack. (See Appendix E, 2009 Fire Management Plan)

Management Direction and Objectives

The actions taken by the Angeles National Forest and the Forest Supervisor with respect to overall incident objectives, controlling the fire at the smallest acreage practicable consistent with firefighter safety considerations, were in concert with the Forest's land management plan. (See Appendix D, Angeles National Forest Land and Resource Management Plan)

Environmental Conditions

The origin of the Station Fire was in extremely rugged terrain with limited opportunities for safe, effective suppression activities. Weather at the time revealed high temperatures with low relative humidity and limited recovery of relative humidity during the evening, leading to unfavorable conditions with respect to the effectiveness of fire suppression. The dense, dry brush was easily ignitable.

Vegetation and Potential Fire Behavior Conditions

The vegetation characteristics in the area of the fire were at very high risk with respect to firefighting operations. Due to the ease of the ignition of the brush, the potential for extreme fire behavior posed unacceptable risk to firefighters. Brush stands were thick and above eye level, obscuring visibility, and making access to the fire difficult.

Initial Attack Response

Firefighters made successful cooperative efforts to engage the fire at critical points during the daylight phase of Initial Attack and made progress toward containing the fire on the portion of the fire above the Angeles Crest Highway (Highway 2). Despite efforts with hose lines and engines, control of the incident was prevented because firefighters were unable to access spot fires that occurred below the Highway at night in steep terrain and in dry, dense brush.

Weather, topography, and dense, dry brush conditions threatened the safety and effectiveness of firefighters the evening of August 27.

Resource Ordering and Assignment

The ordering and assignment of firefighting resources to Initial Attack was appropriate and consistent with best management practices as accepted by wildland firefighting agencies. Additional resources during the evening of August 26 and morning of August 27 would not have improved the effectiveness of operations during that operational period and would have resulted in needless exposure of firefighters to the hazards of wildland fire. Opportunities to affect fire spread were limited on the down slope side of the Highway, not by the amount of suppression resources assigned to the incident, but by the inaccessibility of the fire, hazard of the terrain, and dry, dense brush conditions with extremely limited or no visibility.

Critical Event

On the evening of August 26, spot fires occurred below the Angeles Crest Highway, near the point of fire origin, and were not accessible by firefighters due to excessively steep terrain, limited visibility, and decadent, thick brush. Aircraft use, without subsequent engagement of ground forces, would have been ineffective.

Incident Management Decisions

Based on conditions and circumstances on the afternoon and evening of August 26, 2009, as well as those decisions made subsequently by initial attack incident commanders reflected sound judgment of the operational situation and were prudent with respect to firefighter safety and resource deployment. Their management actions exhibited a consistent awareness of the hazards of the incident and the recognition that engaging the fire below the Highway using ground-based resources would expose firefighters to unmitigated risk. Incident management decisions made during the review period were consistent with broadly accepted incident management practices, including the ordering and assignment of both ground-based and aviation suppression resources.

Conclusion

The purpose of this review was to determine if the Angeles National Forest preparedness actions before the Station Fire were in alignment with existing plans and cooperative agreements; if the decision process and actions taken during initial attack were reasonable and prudent; and to examine the role of the agency administrator in providing leader's intent and input into strategic decisions for the incident initial attack. This review examined only the initial attack phase of the Station Fire from the initial report of the incident on August 26, 2009, to the transition of the incident to the Type I Incident Management Team on August 28, 2009. The review panel concluded that during the initial attack phase of the Station Fire, incident managers on the Angeles National Forest acted in accordance with accepted wildland firefighting practices. They had clear leader's intent and deployed suppression resources only in those conditions where they would be safe and effective. In light of the extremely challenging topography encountered during initial attack, the highly volatile fire, and dense, dry brush, incident commanders were reasonable and prudent in their response and in not exposing firefighters to conditions that would have compromised their safety.

Members of the Review Panel

James E. Hubbard, Team Leader
Deputy Chief for State and Private Forestry,
USDA Forest Service

Tom Harbour, Director
Fire and Aviation Management
USDA Forest Service

John B. Tripp, Chief Deputy
Emergency Operations
Los Angeles County Fire Department

Ken Pimlott, Deputy Director
Chief of Fire Protection
California Department of Forestry and Fire Protection

Donald G. MacGregor, Ph.D.
Senior Scientist
MacGregor Bates, Incorporated

APPENDIX A.

Potential Fire Behavior

Station Fire
Afternoon/Evening of August 26, 2009
David R. Weise, Ph.D.
Supervisory Research Forester

USDA Forest Service
Pacific Southwest Research Station
Forest Fire Laboratory
Riverside, California

General Setting

The terrain below the Angeles Crest Highway near the Angeles Crest Fire Station on the Angeles National Forest is fairly rugged and steep. The slope percentage from the highway to the creek bottom ranges from 33 percent to 67 percent along a 0.5 mile section of the highway centered on the fire station. Several narrow ravines facing east occur along this section of the highway. The slopes below the highway range from 900 to 1,680 feet in length. The vegetation is mature mixed chaparral 6 to 8 feet tall and was at least 50 years old. In contrast, much of the chaparral vegetation on the 2006 Esperanza incident on the San Bernardino National Forest probably ranged from 4 to 6 feet in height. Live fuel moisture is routinely monitored on the Angeles National Forest and in Los Angeles County as a measure of fire danger. Observed live fuel moisture content measured 5 to 7 days prior was 63 percent for chamise, 67 percent for ceanothus, and 89 percent for manzanita. Weather observations at the Little Tujunga weather station (11.4 miles) for 1:00 p.m. included air temperature of 99 ° F, relative humidity of 7 percent, and wind speed of 8 mph. Weather observations at Clear Creek weather station (3.2 miles) included temperature of 89 °F, relative humidity of 11 percent, and wind speed equal to 10 mph. From these observations, predicted fuel moisture of the one hour time lag fuels (easily ignitable grasses, pine needles, small branches) was 1 percent. Moisture content of the larger 10 and 100 hour fuels were 2 and 6 percent, respectively. The spot weather forecast produced at 5:20 p.m., indicated night time temperatures around 74 °F and maximum relative humidity of 17 percent. Maximum observed relative humidity at the two weather stations on the night of August 26 did not exceed 17 percent.

Fuel Conditions

The observed afternoon fuel moisture contents were typical for this time of year; however, the fact that the humidity was not predicted to increase very much on the night of August 26 meant the fuel moistures would stay relatively low throughout the night and that fires would continue to burn actively through the night. Live fuel moisture content was normal for this time of year based on long term means measured by Los Angeles County Fire Department. However, the moisture content in chamise was approaching 60 percent which is considered a critical level for fire behavior. Based on the fire behavior tables found in Appendix B of the Fireline Handbook[5], fuel moisture of the unshaded 1 hour fuels would increase to a maximum of 4 percent by 8:00 a.m. the next morning; however, the equilibrium moisture content associated with 17 percent relative humidity and 74 °F is only 2 percent. The very low moisture content of the fine fuels (1 percent) indicates that these fuels could be easily ignited. This ignition potential was indicated by the forecasted Ignition Component of 100 and the calculated probability of ignition of 1[6]. The fuel conditions described for the Station Fire are similar to conditions associated with the 1968 Canyon Fire fatalities which occurred on August 24, 1968[7]. These investigators hypothesized that a fire whirl occurred near a ridge top and cast a firebrand down slope into a narrow ravine. The resulting upslope fire run lasted 30 seconds and overtook the crew. The fuel age where these fatalities occurred was estimated to be 50 to 80 years.

Anticipated Fire Behavior

Several options are available to estimate the potential fire behavior that the initial attack firefighters would have experienced from a fire that started in the chaparral below the Angeles Crest highway. All of these options are based on the Rothermel fire spread model[8]. Most of the versions describe chaparral as 6 feet deep. Cohen developed a variant called FIRECAST[9] that contained fuel models specific to chamise and mixed chaparral of various heights and ages. I used FIRECAST to estimate the flame length, fireline intensity, and rate of spread for the forecasted night time conditions (dead fuel moistures: 1 hour equal to 1, 10 hour equal to 2, Slope equal to 50 percent, midflame wind speed of 6, 8, and 10 mph, live fuel moisture of 63 percent, and percentage of fuel bed that was dead of 30 and 50 percent). Predicted spread rates for these conditions ranged 0.75 to 1.5 mph, flame length ranged from 18 to 30 feet and fireline intensity ranged from 2,900 to 8,400 BTU/ft/s. The spread rates and flame lengths are similar to the fire behavior anticipated by the fire danger indices calculated for the Clear Creek and Little Tujunga stations using the weather data collected at 1:00 p.m., on August 26, 2009.

Terrain Effects

The area below the highway is relatively steep and thus difficult to walk up. The slopes in the 1948 Mann Gulch Fire[10] fatalities and the 1994 South Canyon Fire[11] fatalities were 44 to 76 percent and 55 to 80 percent, respectively. In the 1968 Canyon Fire which occurred 17 miles east of the Station Fire

Night Time Fire Behavior

The fire behavior predicted to occur during the evening and night was active primarily for one reason, no humidity recovery. Typically at night, temperatures drop and relative humidity increases resulting in an increase in fuel moisture. The increased fuel moisture causes fire behavior to moderate. However, relative humidity was not forecast to increase appreciably due to the nature of the air mass over the Station Fire. Fuel moistures stayed low and fire behavior did not moderate. Spot fires could have spread successfully at night since fine fuel moisture content was low.

Interpretation of Fire Behavior

Guidelines for suppression activities that are appropriate for different levels of fire behavior have been developed over the years. Flame length, rate of spread, and fire intensity are the fire behavior indicators typically used and can be displayed on a fire characteristics chart. The predicted fire behavior was plotted on the attached fire characteristics chart. Note that the total amount of energy released did not change as the spread rate changed. The * indicates that chaparral with 30 percent dead material and the * indicates chaparral with 50 percent dead material. Fireline intensity describes the rate at which the energy is released which has been correlated with flame length. Fireline intensities < 100 BTU/ft/s or flame lengths < 4 ft can be directly attacked by firefighters with hand tools. Flame lengths 5 to 8 feet generally require some sort of ground-based vehicle attack (engine, bulldozer). For the Station Fire, the predicted fire behavior far exceeds any kind of firefighting capabilities that ground-based resources might have.

Summary

Predicted fire behavior potentially encountered by initial attack forces on the Station Fire was typical for August. This fire behavior is extreme and not amenable to attack by hand crews or fire engines. Predicted fuel moisture conditions had a high probability of supporting easy ignition of fine fuels by spot fires, fast spread rates, and large flames. Terrain was very steep and difficult to traverse on foot, which has been typical of areas where firefighter fatalities have occurred.

Interpretation of Fire Behavior

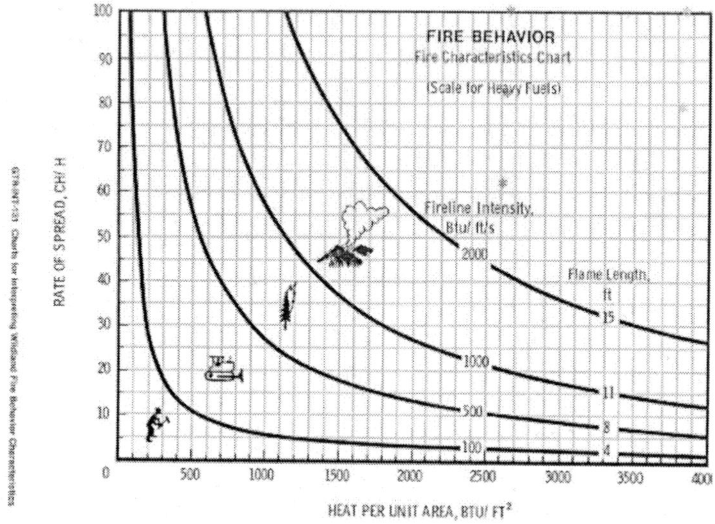

Predicted fire characteristics associated with potential spot fires occurring below the Angeles Crest Highway, near Angeles Crest Fire Station, August 26, 2009. Red stars are for fuel beds with 30 percent dead material, orange are for 50 percent dead materials.

APPENDIX B.

Station Fire: Photographs of Area of Initial Attack

Photograph 1. Small, unnamed spur canyon (center) within Woodwardia Canyon, location of the initial perimeter of the Station Fire on August 26, 2009. Point-of-origin located near dirt turnout along Highway 2 in foreground. Photo taken from Civilian Conservation Corps Ridge, location of the original Incident Command Post, looking west-southwest. (Angeles National Forest, Los Angeles River Ranger District. Photo taken on October 21, 2009, after first significant post-fire rainfall. Strong red coloration from retardant somewhat diminished from upper canyon and side ridges.)

Photograph 3. Looking north along S.R. 2 (Angeles Crest Highway). Southern edge of original Station Fire perimeter on ridge to left. During the early morning hours of August 27, 2009, down canyon winds carried embers from this ridge to unburned vegetation on the downhill (east) side of S.R. 2 to the right. Photo taken on October 21, 2009.

Photograph 4. Looking east-southeast across S.R. 2 (Angeles Crest Highway) from within the original Station Fire perimeter. From this location, down canyon winds carried embers across S.R. 2 (east) into the vegetation below. Photo taken on October 21, 2009.

Photograph 5. Looking upslope to the west into the initial footprint of the Station Fire. Photo taken along S.R. 2 (Angeles Crest Highway) from turnout near where the fire began on August 26, 2009. Photo taken on October 21, 2009.

Photograph 6. Steep slopes with nearly impenetrable chaparral vegetation within the initial footprint of the Station Fire. Photo taken looking west from along S.R. 2 (Angeles Crest Highway) near where the fire began on August 26, 2009. Photo taken on October 21, 2009.

APPENDIX C.

Station Fire Initial Attack

Area of Initial Attack and Fire Perimeter (Google Earth Representation)

Station Fire—August 26, 2009

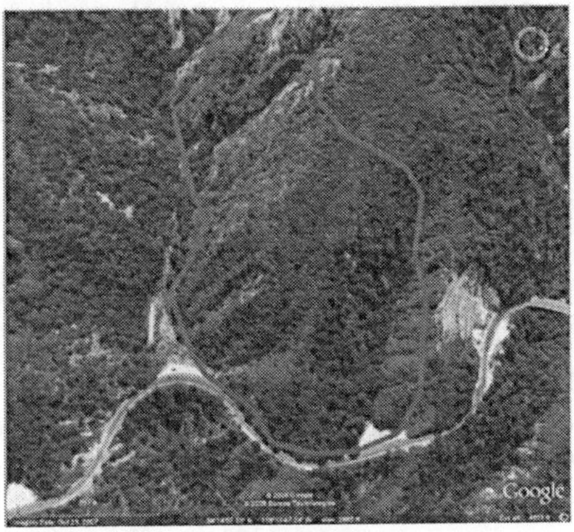

Station Fire Initial Attack Area of Initial Attack and Fire Perimeter (Google Earth Representation)

Station Fire—August 26, 2009

Station Fire Initial Attack
Area of Initial Attack, Fire Perimeter, and Point of Origin
(Google Earth Representation)
Station Fire—August 26, 2009

Photograph 2. Small, unnamed spur canyon (center) within Woodwardia Canyon, location of the initial perimeter of the Station Fire on August 26, 2009. Area to lower-left of center (exposed highway cut) is where the Station Fire spotted east (down canyon) across Highway 2 during the early morning hours of August 27, 2009. Photo taken from Civilian Conservation Corps Ridge, location of the original Incident Command Post, looking west-southwest. (Angeles National Forest, Los Angeles River Ranger District. Note: Photo taken on October 21, 2009, after first significant post-fire rainfall. Strong red coloration from retardant somewhat diminished from upper canyon and side ridges.)

Station Fire Progression Map
August 26—August 28, 2009

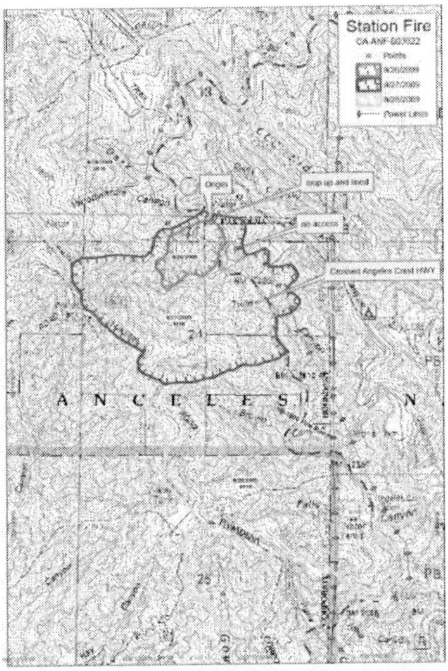

Google Earth map demonstrating where fire cross the Angeles Crest Highway, the spot fire below the road, point of origin and where fire was lined.

Alternate view of fire origin, spot fire and where the Station Fire crossed the Angeles Crest Highway.

Alternate view of origin,spot fire and where the Station Fire crossed the Angeles Crest Highway with fire progression lines Representing the fire perimeter August 26- 28, 2009.

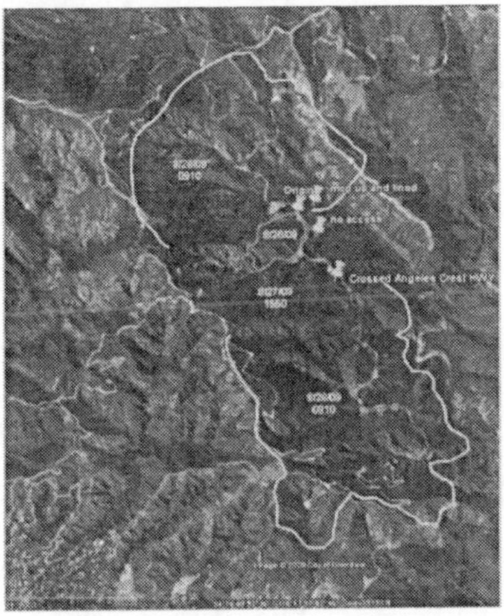

Fire Progression Map from August 26 through August 28, 2009, at 9:10 a.m.

APPENDIX D.

Excerpt from the Angeles National Forest 2005 Land Management Plan Direction for Fire Management and Fire Suppression

USDA

United States Department of Agriculture

Forest Service

Pacific Southwest Region

September 2005

Land Management Plan

Part 2 Angeles National Forest Strategy

Page 1, Angeles NF Forest Plan Revision Record of Decision

"This revised forest plan meets our legal obligations to the people of the United States and the national forest environment that surrounds them. The Forest Service understands its role or niche within the network of communities throughout southern California. The Angeles National Forest is a unique regional feature with important resources that must be sustained over time. I believe that my decision maximizes the net public benefit (sustainability of social, economic, and ecological processes) to meet current needs and to leave options for future generations to continue to enjoy sustainable recreation opportunities, healthy forest systems and appropriate community protection from wildland fire. These benefits are achieved through proven measures that protect, maintain, improve, and restore the health of the national forests and open lands; reduce risks from wildland fire, invasive species, insects, disease and other threats; provide wildlife habitat; protect and restore unique vegetation and terrain; and perhaps most importantly, to provide a diversity of recreation, environmental education opportunities, and monitoring requirements, that are the cornerstone of our program emphasis for the future."

Page 4, Angeles NF Forest Plan Revision Record of Decision

Chief's Four Threats

The revised forest plan is responsive to the Forest Service's 2004 National Strategic Plan and to the resolution of the four threats described by the Chief of the Forest Service (Publication Speech, 2003). These four threats include:

- **Fire and Fuels** – decades of fuel buildup, coupled with drought and disease, have created a situation that poses a real threat to the lives and property of people living in the communities of southern California. In southern California, fire is a fact; it is not a question if fires will burn, rather, it is a question of when and how intensive.

Page 6-7, Angeles NF Forest Plan Revision Record of Decision

Developed Area Interface: Much of the urban interface and concentrated use areas within the Forest boundary are included in the Developed Area Interface zone. Accordingly, much of our community defense work and fuels management activities in response to the National Fire Plan will be focused within this zone. The Developed Area Interface (DAI) zone includes the Wildland/Urban Interface (WUI) zone described in the National Fire Plan. Within the WUI zone there are two sub-zones called the WUI Defense zone and the WUI Threat zone. Areas beyond these zones can, under the right conditions, be included in the WUI zone and therefore are described

Page 6 Angeles NF Forest Plan Revision Record of Decision

As the WUI influence zone. The DAI zone is especially compatible with community defense work and already includes a significant amount of the National Forest System roads that enable access for community defense work or fire suppression. The zone is also compatible with the location of sites needed for a variety of special uses.

Page 13, Angeles NF Forest Plan Revision Record of Decision

Alternative 4a (Selected)

Alternative 4a adjusts the preferred alternatives by using selected elements from other alternatives, as well as making changes to the scheme of land use zones in response to public comment, tribal consultation, and internal review of the draft environmental impact statement (DEIS) and land management plans for the four southern California national forests.

Alternative 4a is focused on active management for the maintenance of healthy forests; community protection from wildland fire; managed, sustainable recreation settings and uses; and the management of threatened and endangered species. The alternative theme includes the opportunity for a low level of growth of recreation activities and the facilities to support increased use. Managed sustainable use of the national forests is compatible with the maintenance of long-term biological diversity and ecological integrity. The focus on community protection is complementary to the National Fire Management Policy.

Page 15, Angeles NF Forest Plan Revision Record of Decision

Given these criteria, I am identifying Alternative 4a as the environmentally preferred alternative. This Record of Decision has discussed the decision process and the rationale for the decision. Given the complexities of the management situation in southern California and the way Alternative 4a addresses the conservation and protection of plant and animal species and their habitat, the demand for human uses, and the critical need for community wildland fire protection, it just makes sense. The evaluation process that I have described in the previous section includes the evaluation of net public benefit, the key factors, and the attributes and advantages that cause Alternative 4a to stand out, in my mind, as environmentally preferable.

Pages 32-34 LMP Part 2: Angeles National Forest Strategy

Fire Management

Fire Management includes all activities involved with pre-fire preparation, fire hazard reduction such as brush removal, and public education concerning fire prevention and safety.

The Fire Management and Administration group formulates and administers fire management and safety plans, and oversees all fire management operations including budget and planning, general supervision, scheduling, and other administrative activities.

Wildland fire suppression 2005 encompasses all activities included in containing and mitigating the damages of wildland fires caused by either natural or human means. This program also includes national support of fire and disaster teams in other areas of the country. The primary responsibility is in supporting large suppression operations nationally; however, other types of assignments include assisting the Federal Emergency Management Agency (FEMA). Past assignments have included earthquakes, floods, hurricanes, 9/11 disaster support, and supervision of the Columbia Space Shuttle debris recovery.

Prevention is based on three primary categories: education, engineering and enforcement. Education includes Smokey Bear programs to instill a fire prevention ethic in school children and Firewise community programs that target civic and homeowner groups. Engineering includes abatement of fire hazard along roadways and in high-use areas using fire retardants and removal of flammable vegetation. Enforcement includes executing state fire law regarding hazard abatement around structures, for both public and private land in the national forest. This is also done along all electrical transmission and distribution systems, (placed by public utility agencies), across the national forest.

Hazardous fuel reduction is the set of activities associated with removing brush and vegetation from areas where they pose a significant threat to human life, property, and national forest resources, and where they interfere with the health of natural fire-adapted ecosystems. Fuel reduction involves direct management of vegetation using prescribed fire, mechanical, manual, or chemical methods. This is accomplished by a multidisciplinary planning approach using resource specialists, local governments, communities and contractors. The national forest Fuels Officer provides overall leadership for this program, which is then carried out by Fire Management personnel and local government.

Suppression of wildland fires is the first priority for program managers. All wildland fires on southern California national forests are considered to be a threat to communities. Aggressive fire suppression and prevention strategies will be implemented near communities to achieve the objectives to protect life and property from wildland fire, subsequent floods and debris flows. National Forest staff plan to maintain the suppression organization at 90 percent of the most efficient level or higher.

WUI Defense and Threat Zones around structures, fuelbreaks, and vegetation treatments to maintain or restore forest health within community protection areas are the next priority. Over the next three to five years, vegetative treatments will be strategically integrated to maximize community protection efforts and minimize wildland fire size, while considering habitat needs. Mortality removal will be integrated with thinning within the community protection areas. National Forest staff plan to complete approximately 40 percent of identified treatment needs (see Fire 1 – Fire Prevention; Fire 2 - Direct Community Protection; Fire 3 - Fire Suppression Emphasis; Fire 4 - Firefighter and Public Safety; Fire 5 - Fuelbreaks and Indirect Community Protection; and FH 3 - Restoration of Forest Health).

Pages 124-128, LMP Part 2: Angeles National Forest Strategy

Fire 1 - Fire Prevention

Reduce the number of human-caused wildland fires and associated human and environmental impacts. Focus fire prevention programs on the urban interface, threatened, endangered, proposed, candidate and sensitive species habitat, vegetative areas threatened with type conversion and areas of major recreation use:

Continue environmental and fire prevention classroom education in local schools to reach the diverse communities the Forest Service serves.

Implement Forest Fire Restrictions and Closure Plan as appropriate, including an internal/ external public communication plan.

Continue the activation and utilization of the National Fire Prevention and Education Teams as appropriate in order to augment local fire prevention resources.

Develop and implement a plan to expand structure hazard reduction from 30' zones to 100' zones of clearance.

Continue to refine the process of implementing partial or full national forest closure as appropriate in order to increase the margin of public and firefighter safety.

Linked to National Strategic Plan Goal 1- Reduce the risk from catastrophic wildland fire, objective 2.

Fire 2 - Direct Community Protection

Reduce the number of high risk/high value, and high and moderate risk acres using mechanical treatments, grazing, and prescribed fire. Identify and schedule for treatment the high risk and high value acres near communities and developed recreation sites, including the installation of Wildland/ Urban interface (WUI) Defense and Threat Zone vegetation treatments. Highest priority should be given to those areas with substantial drought and insect-killed vegetation that present a significant threat to life and property in entire communities:

Promote removal of tree mortality adjacent to structures as the first step in reducing threats to human life and investments.

When National Forest System lands are managed for direct community protection, consider the use of Memorandums of Understanding with Fire Safe Councils as a means of allowing residents to meet state fire law or county brush clearance ordinances on a combination of private and public lands.

Herbicides or the repetitive use of prescribed fire may be used in the WUI Defense zone on National Forest System land to avoid expensive treatments of resprouting chaparral species.

Linked to National Strategic Plan Goal 1- Reduce the risk from catastrophic wildland fire, objectives 1 and 3.

Fire 3 - Fire Suppression Emphasis

All fires either on the national forest or that threaten the national forest will be suppressed. Wildland fires that pose life safety threats, threaten communities, improvements or infrastructure will receive a management response commensurate with minimizing acres burned. An appropriate management response (including a full range of suppression strategies) may be used elsewhere on the national forest where natural and cultural resource impacts along with fire suppression costs are primary concerns.

Cross train with other fire agencies to improve suppression coordination and performance on fires burning in the Wildland/Urban Interface or developed area intermix.

Continue to expand mutual aid agreements with fire cooperators in order to ensure the coverage of key fire stations during periods of fire resource drawdown.

Continue the evaluation of current and future fire station locations with respect to strategic location, changing demographics, suitability and operating costs.

Implement a "Fireshed" analysis of the national forest to either validate or modify current fire management strategies and tactics.

Linked to National Strategic Plan Goal 1- Reduce the risk from catastrophic wildland fire, objective 2.

Fire 4 - Firefighter and Public Safety

Improving firefighter and public safety is the primary objective in fire management. All other activities are tiered to this core value. Integrate fire management activities with those of other government agencies that share a mutual interest or benefit in fire activities on the national forest. Conduct fire management activities in a cost effective manner:

Improve residential inspection capability to enhance the defensible space around structures.

In concert with other agencies and Fire Safe Councils develop evacuation and structure protection plans that will enhance both firefighter and public safety.

Maintain program reviews, training and qualification standards contained in the Fire Management Plan.

Coordinate meetings, training and workshops with employees and cooperating fire agencies to improve fire and emergency operations capability and preparedness.

Cooperate with local agencies to develop and build-out an inter-operable radio communications system for emergency response and incident management in southern California.

Linked to National Strategic Plan Goal 1- Reduce the risk from catastrophic wildland fire, objective 2.

Fire 5 - Fuelbreaks and Indirect Community Protection

Maintain the existing system of roadside fuelbreaks and fuelbreaks along watershed boundaries to minimize fire size and the number of communities threatened by both fires and floods. Consider constructing new fuelbreaks on land outside of wilderness or other special designations.

Consider an opportunistic approach to fuels management. Take advantage of previously burned or treated areas to link future fuels and vegetation management or wildlife habitat improvement projects when doing fuels planning.

Utilize fire landscape analysis to aid in the design of future fuelbreak systems, maintain multiple lines of community defense, and to minimize future wildland fire size.

Develop a plan to minimize the propagation of invasive nonnative species during fire suppression and fuels or vegetation management activities.

Linked to National Strategic Plan Goal 1- Reduce the risk from catastrophic wildland fire, objectives 1 and 3.

APPENDIX E.

Excerpts from the Angeles National Forest
2009 Fire Management Plan
Fire Management Unit 2
(Area of the Station Fire)

ANGELES NATIONAL FOREST
2009 Fire Management Plan

Prepared by: ______________________ Date: ________
David Kerr
Adaptive Management Services

Reviewed by: ______________________ Date: ________
John Thomas
Deputy Chief, Fire and Aviation Mgmt.

Reviewed by: ______________________ Date: ________
David Conklin
Chief, Fire and Aviation Management

Approved by: ______________________ Date: ________
Jody Noiron
Forest Supervisor, Angeles National Forest

Fire Management Unit – 2 – Mid-Elevation, Non-Wilderness

Fire Management Unit Characteristics

This Fire Management Unit is characterized by a mid-elevation shrubland system and includes limited in-holdings, reflecting a true wildland urban intermix fire environment. Scattered developments including Forest Service facilities, organizational camps, electronic sites and recreation developments add to the complexity of the fire situation. The density of fire suppression resources has decreased significantly from FUM 1, as cooperating resources no longer are available through automatic aid.

The FMU is best characterized by fuel type and fire behavior. The density of the fuelbed and canopy closure increases significant in this mid-elevation range. Grasslands sages fall out of the stand structure and a transition to woody chaparral vegetation occurs. Fuel loads increase to 16+ tons per acres (Weise and Regelbrugge, 1991) on many sites with an associated increase in fireline intensity and resistance to control. Aerial firefighting resources are less effective due to the closure of the canopy and increased fire intensity. Production rates decreases for crews and engines in this FMU as compared to FMU-1.

The standard initial attack dispatch to a report of a wildland fire is the same as FMU- 1.

Topography and vegetation vary greatly within the FMU. Large portions of the FMU were burned during the Copper, San Fran, Bouquet, Williams and Curve fires of 2002. These fires were stand-replacing incidents and have modified the fuel component in major watersheds. The FMU includes the primary portions of the major watersheds for the Forest. These include San Gabriel Canyon (West, Middle and East Forks), Arroyo Seco Canyon, Big Tujunga Canyon, Little Tujunga Canyon, Pacoima Canyon, Bouquet Canyon, Lake Hughes Canyon and San Fransiquito Canyon. These primary watersheds and their heavily bisected nature represent the major topographic features that effect fire behavior and local weather patterns.

The lack of fire frequency in this mid-elevation range has left a plant community that is relatively intact, resilient and vigorous. While recent fires have affected large portions of the FMU, large expanses of continuous chaparral remain.

Aspect plays a major role in the distribution of chaparral species. North facing slopes include a live oak component that was missing from FMU- 1. several species of Ceanothus begin to appear in the plant community and Manzanita also became a conspicuous element of the fueled. Chamise, lemonade berry, toyon and Laurel Sumac continue to be important species. Coulter pine at the upper elevation range of the FMY and Big-cone Douglas-fir are the coniferous species found. Big-cone Douglas-fir has been identified as a species of special concern in the *Southern California Mountains and Foothill Assessment*, since it has been identified as falling out of the ecosystem due to increasing fire intensity in and adjacent to the stands. This increase in intensity is associated with fire exclusion within the stands (Minnich, 1999).

The following is the Critical Biological (CB) Areas and the primary species protected (Land Management Plan Part 2, page 11):

- **Castaic** (367 acres) – Arroyo toad
- **Fish Canyon** (104 acres) – Arroyo toad
- **San Francisquito Canyon** (130 acres) - California red-legged frog, unarmored three-spine stickleback and Berberis nevinii (Nevin's Barberry)
- **South Fork Little Rock Creek** (12 acres) – Mountain yellow-legged frog
- **Upper Big Tujunga** (818 acres) – Arroyo toad, California red-legged frog
- **West Fork San Gabriel River** (506 acres) – Santa Ana sucker
- **Lower Little Rock Creek** (86 acres) – Arroyo toad

Incident managers need to be aware of the presence of these species and the potential effects of suppression operations on habitat areas. Direction issued from the United States Fish and Wildlife Service in 1995 indicates that firefighter and public safety will not be constrained in order to protect endangered species or their habitats.

In order to mitigate issues concerning TESPC species and other resource concerns, a Resource Advisor from the Forest will be ordered on all fires where aerial retardant use has or is likely to occur in water courses or in areas of modeled/occupied habitat for threatened and endangered species. The IC is also responsible for documenting actions that occurred within known occupied and modeled habitat areas so that appropriate mitigation actions can be initiated after the incident.

Table 3.3. Acres by Land Use Zone in FMU 2

FMU	BC	BCMUR	BCNM	CB	DAI	EF	EW	RW	Grand Total
2	91,677	26,563	124,151	2,023	39,621	14,368		10,011	296,682

Historic and cultural resources are presence within the FMU and tend to be clustered adjacent to the broad valley bottoms, watercourses, Valley and Coast Live Oak clusters and saddles and ridgelines. The Forest archeologist maintains records of known artifact locations and can brief on-scene fire managers concerning these resources. The sensitive nature of the information related to heritage resources do not allow for full dissemination of this information to all module leaders on the Forest; however this information is maintained by the Chief Officers.

Real property abounds within the FMU and a comprehensive list of these properties is not useful. Pre-attach maps maintained by the Forest identify key facilities and communities within the FMU.

Strategies and Measurable Management Objectives

The resilience of the established shrub lands within the FMU make this Unit a target for the use of prescribed fire and other vegetation management treatments. Key suppression control features, such as the Santa Clara and Sierra Pelona fuelbreaks are located primarily within the boundary of the FMU. These features need to be maintained through the use of prescribed fire, mechanical or biological means.

Fire suppression will continue to focus on aggressive initial attack in attempts to minimize acres burned during wildfire. This strategy provides the best opportunity to protect resource management features such as Bigcone Douglas-fir and TESPC species habitat while protecting improvements and private in holdings aggressive suppression also helps prevent fire from burning out of the interior of the Forest into the adjacent wild land urban interface.

Fire prevention activities and mitigation work will assist in developing and maintaining defensible space around improvements, with prevention patrols focused on areas of high public use or clusters of historic enhance tactical fire suppression opportunities.

Objectives are:

- Provide an appropriate management response to reported wildland fires within the FMU. Maintain a minimum initial attack capability year round to assure an appropriate management response is provided.
- Assure a minimum of one qualified duty office is available for immediate dispatch to provide incident oversight on a year round basis. An on-call Chief Officer is identified for each Ranger District and a Forest Fire Staff Officer is available to respond to a reported incident 24-hour a day.

- Assure a Wildland Fire Situation Analysis is completed for all fires which are uncontrolled after the second burning period.
- Assure that wildland fires are evaluated for appropriate management oversight using the *IRPG and "Redbook" Complexity Analysis Guide*.
- Conduct prescribed fires within established prescription windows and with only fully qualified personnel.
- Coordinate prescribed fire activities with the appropriate air quality regulators.
- Complete fire hazard inspections for Forest Service facilities prior to June 15 annually.
- Complete fire hazard abatement work prior to July 1 for Forest Service facilities.

Management Constraints Affecting Operational Implementation

Management constraints in the area of fire suppression and fire prevention are minimal. Activities need to be closely coordinated with agency resource managers to assure that sensitive resources are protected to the greatest extent possibly without comprising firefighter and public safety or unduly placing private property at risk during fire events. The area of prescribed fire has identified operational constraints. These constraints include:

The South Coast Air Quality Management District is a severe non-attainment area under the National Ambient Air Quality Standards. The degraded nature of the air shed limits the number of allowable burn days constraining prescribed fire opportunities.

The presence of modeled and occupied habitat for TESPC species requires close coordination with the United States Fish and Wildlife Service to obtain an affects determination for proposed treatments. This coordination and any mitigation actions required constrain treatment areas and burn windows.

Historic Fire Occurrence

While this FMU has a relative low number fire starts annually, it has the potential for large fire development given the nature of the fuel structure and the completely of the terrain. Fires burning our of the wildland urban interface and into the Forest have heavily impacted the FMU. Flanking tactics utilized on fires starting in the lower elevations tends to push fire spread into this unit. Ignition patterns near transportation corridors. Lightning fires occur in the upper elevations of the FMU, however most lightning activity is concentrated in FMU-3 and FMU-4.

The Fire Management Situation

Weather patterns influencing fire behavior and historical weather analysis

The FMU has two unique weather patterns that influence fire behavior and activity. The desert side of the FMU as defined by Newhall Pass north through the Santa Clarita Valley and the Antelope Valley portion of the Unit. This area trails FMU-1 in supporting active fire spread and ignitions. The desert influenced areas experience strong winds in the late spring and summer as the thermal low develops over the Mojave Desert. The pressure gradient between cold marine air and the hot desert sets up strong circulation patterns that leads to gusty and highly variable winds. This desert-influenced portion of the FMU also shows the earliest curing of annual grasses and has the lowest average live fuel moistures.

While the same weather pattern impacts the front range of the FMU it is expressed differently, as the air mass still contains much of the moisture associated with the coastal zone. This can lead to low clouds and fog through the lower elevations, however this marine influence normally establishes itself below 3,500'. The fire environment above this inversion is significantly different. Fire behavior and suppression decisions can vary based on how deep the inversion has established. Drying trends for live and dead fuels as well as wind patterns are different based on this weather feature.

As the pressure gradient associated with the thermal low decreases during the summer, weather conditions between the desert influenced and front range portions of the FMU become more closely aligned.

Automated weather stations most reflective of the FMU are Tanbark, Warm Springs, Clear Creek and Camp 9. Camp 9 is operated by the Los Angeles County Fire Department.

Fire season determination

Fire activity in this FMU rotates around the Unit as the annual grasses cure and live fuel moisture drop. The annual grasses act as the wick to spread the ignition source into the live fuels. Curing dates vary annually, but can be as early as April 15, dependent on the amount and distribution of the winter rains.

Live fuel moisture sampling within the FMU occurs at Tanbark, Clear Creek, Little Tujunga and Coarse Gold. These samples are taken on a bi-weekly basis during the months of May through December. Monthly samples are taken the remainder of the year. This information is consolidated at the Angeles ECC and is available to Chief Officers as a decision making tool.

Fuel conditions likely to affect fire behavior

Some dieback of chaparral species occurred during the summer of 2002 as moisture stress impacted the vegetation. Manzanita is the dominant species impacted by the dieback, but some species of Ceanothus also display die back. This event is not wide spread and represents a localized change to the fire environment. Coulter pine, especially plantations at lower elevation display significant mortality.

From a fuels perspective fire behavior is a factor of live and dead fuel moisture, curing date and dead fuel loading. Cured fine fuels in this FMU are not generally continuous enough to cause fire spread at relatively high live fuels moistures. As the season progresses and live fuel moisture drops fire activity increases as the live fuels contribute energy into the fire environment.

Fire regime alteration

The fire regime associated with this vegetation type is high intensity, stand replacing in nature. The Fire Return Interval (FRI) is classified as 35 to 100 years. This return interval is contradicted by information found in Southern California Mountains and Foothills Assessment, which identifies the FRI as 50 to 80 years (Stevenson, Calcarone, 1999).

Overall the chaparral plant community is vigorous and resilient. Fire frequency is such that seed banks have been able to reestablish after fire and remain a viable source of recruitment for obligate seeding species. Spouting species continue to display viability after fire.

Control problems and dominate topographic features

Steep slopes and heavily bisected topography limit control opportunities within the FMU. Mechanized equipment is restricted in many areas to the primary ridge systems. Fires under typical weather patterns run to the ridge tops where changes in the alignment of the fire spread allow for successful suppression operations to be conducted. Lateral spread is a function of winds and fuel moistures. As live fuel moisture decreases, lateral spread increases in the chaparral fuel type, this lateral spread often allows a fire to reposition for rapid upslope fire runs.

The density of the chaparral fuels, especially fuel beds older than 20 years, reduces the effectiveness of aerial suppression actions as retardant and water is less efficient at penetrating the canopy and affecting the surface fuels.

Other elements of the fire environment affecting fire management

The wildland urban intermix consisting of private property, structures and federal facilities represent the greatest values to be protected. Resource values focus primarily on protection of riparian habitat, soil stabilization and water quality. Requirements for the protection of riparian habitats are found the LMP, Part 3, Appendix E and F.

The density of roads within this FMU decreases when compared to FMU-1. Lack of access is a limiting factor for fire control resources. The ability to maintain roads to *"provide safe and efficient routes for ... fire protection equipment and other high clearance vehicles"* (LMPR, Part 2, Trans 1) is an important fire suppression requirement.

End Notes

[1] For the purpose of this review the term "initial attack" encompasses the actions of the incident commanders from the initial dispatch on August 26, 2009, through to the transition of the fire to the Type 1 Incident Management Team on August 28, 2009, at 2:00 p.m.

[2] Haimes, Y.Y. (1998). Risk modeling, assessment and management. New York: Wiley.

[3] National Research Council (2009). Science and decisions: Advancing risk assessment. Washington, D. C.: The National Academies Press

[4] Resources that initially responded from Angeles National Forest and LA County Fire Department on August 26, 2009 included: Nine (9) handcrews of 20 members each, 2 Chief Officers from the Angeles National Forest and 2 from LA County Fire, 13 engines, 3 water tenders, 2 patrols, 2 medium helicopters, 1 heavy helicopter, 1 heavy Helitanker, and an Airco helicopter. Additionally, the Morris Fire sent 1 air attack, 2 Helitankers and 2 airtankers.

[5] National Wildfire Coordinating Group (NWCG). 2006. NWCG Fireline Handbook, Appendix B; Fire behavior. PMS 410-2, NFES 2165. Boise, ID

[6] Bradshaw, L.S.; Deeming, J.E.; Burgan, R.E.; Cohen, J.D., compilers. 1984. The 1978 National Fire Danger Rating System: technical documentation. Gen. Tech. Rep. INT-169. Ogden, UT: U.S. Department of Agriculture, Forest Service, Intermountain Forest and Range Experiment Station. 44 p.

[7] Countryman, C.M., McCutchan, M.H., Ryan, B.C. 1969. Fire weather and fire behavior at the 1968 Canyon Fire. Res. Pap. PSW-55. Berkeley, CA: U.S. Department of Agriculture, Forest Service, Pacific Southwest Forest and Range Experiment Station. 20 p.

[8] Rothermel, R.C. 1972. A mathematical model for predicting fire spread in wildland fuels. Res. Pap. INT-115, Ogden, UT: U.S. Department of Agriculture, Forest Service, Intermountain Forest and Range Experiment Station. 40 p.

[9] Cohen, J. D. 1986. Estimating fire behavior with FIRECAST: user's manual. Gen. Tech. Rep. PSW-90. Berkeley, CA: U.S. Department of Agriculture, Forest Service, Pacific Southwest Forest and Range Experiment Station. 11 p.

[10] Rothermel, R. C. 1993. Mann Gulch fire: a race that couldn't be won. Gen. Tech. Rep. INT-299. Ogden, UT: U.S. Department of Agriculture, Forest Service, Intermountain Research Station. 10 p.

[11] Butler, B.W.; Bartlette, R.A.; Bradshaw, L.S.; Cohen, J.D.; Andrews, P.L.; Putnam, T.; Mangan, R. J. 1998. Fire behavior associated with the 1994 South Canyon Fire on Storm King Mountain, Colorado. Res. Pap. RMRS-RP-9. Ogden, UT: U.S. Department of Agriculture, Forest Service, Rocky Mountain Research Station. 82 p.

[12] Andrews, P.L.; Rothermel, R. C. 1981. Charts for interpreting wildland fire behavior characteristics. Gen. Tech. Rep. INT-131. Ogden, UT: U.S. Department of Agriculture, Forest Service, Intermountain Forest and Range Experiment Station. 21 p.

[13] Byram, G.M. 1959. Forest fire behavior. Pp. 90-123 In: Forest fire: control and use. Edited by K.P. Davis. McGraw-Hill, New York.

In: The Station Fire ISBN: 978-1-62081-084-2
Editors: E. L. Donahue and D. M. Elmore© 2012 Nova Science Publishers, Inc.

Chapter 3

STATION FIRE LESSONS LEARNED REPORT*

USDA Forest Service

NIGHT FLYING OF AIRCRAFT

The Forest Service has received criticism for not utilizing night flying capability by partnering agencies during the evening of August 26 and early morning hours of the 27th , 2009. As a result, the Forest Service identified this topic as an area to explore for potential use in the future.

National

The Chief of the Forest Service has initiated a national assessment of night flying operations. The Forest Service stopped these operations in the early 1980's due to night flying accidents that resulted in fatalities. So far, the preliminary findings and options for rotor-wing aircraft are under review. Fixed wing night flying finding will be assessed during the winter of 2010 and 2011.

- The study identified Kern County, Los Angeles City, Los Angeles County, Orange County, San Diego City, and Santa Barbara County

* This is an edited, reformatted and augmented version of the USDA Forest Service, dated October 2010.

are currently or preparing to operate helicopters in a night operations mode.

- There are an estimated 17 helicopters that support the current night operations in Southern California and their primary operational missions are emergency medical service.
- Utilization of the aircraft has been minimal for actual firefighting night operations and their zone of coverage is well defined. This is to ensure the safety of the pilot and aircraft so the familiarity of the terrain and any hazards are well known in advance of flying a mission.
- CALFIRE has chosen not to develop this capability at this time due to the age and characteristics of the helicopters they operate.

Elements of the Study: The study reviewed the past practice of using helicopters in support of night firefighting, which agencies are currently doing night operations, what technology is available today, mission definition and qualification, risk assessment for the candidate helicopter night missions, effectiveness and utilization of night water drops and cost estimates. The missions that were identified to move forward for further evaluation were:

- Water and retardant dropping using a fixed tank with ground fill
- Aerial supervision
- Emergency medical transport (Hoist)
- Aerial ignition with plastic sphere dispenser

Risk Assessment: The assessment utilized the 2008 Systems Safety Aviation Guide and the 2009 Aviation Risk Management Workbook, and identified all hazards to be classified as Section 5 (c), FAA Circular 120-92, acceptable with mitigation. There were five systems identified with an additional 24 subsystems following the workbook. The five systems are:

- Helicopter Aircraft;
- Helicopter Facilities Night;
- Helicopter Personnel Night;
- Helicopter Technology Night; and
- Helicopter Operations Night.

Regional

The Region will explore and develop training for utilizing cooperators night flying operations for all Incident Management Teams in the Pacific Southwest Region.

Develop night flying operations briefing package for all out-of-area resources to review, including Incident Management Teams.

Augment existing aviation management training to include management of night flying operations.

Local- Angeles National Forest (ANF)

The Angeles NF Forest Supervisor and LA County have agreed to modify the Cooperative Agreement Annual Operating Plan (AOP) to specify management's expectations regarding night flying. Routine training and operations will ensure Incident Commanders (IC) are aware that they may approve the use of properly fielded night-flight helicopters during evening and early morning hours to contain a fire that is an immediate threat or may become a threat to life and property, when appropriate, and or when both agencies are in Unified Command. This would apply to all Angeles NF lands, including lands within the Initial Action Zones. Angeles NF personnel will receive additional training on safety requirements when using night flying helicopters. The Forest will pay for these operations as 'Assistance by Hire'.

At the 2010 spring forest chief officer meeting, and annual spring chief officer meeting with Los Angeles County, the topic of night flying operations was discussed. All forest chief officers with duty officer status are aware that they **can** utilize Los Angeles County Fire helicopters for night flying fire suppression support anywhere on the Angeles National Forest.

Fire Behavior Modeling Improvements

Fire behavior during the Station Fire was unprecedented and proved difficult to model. It has been recognized that there is a need to enhance and update the fire behavior modeling tools to better support decision making in the complex Southern California fire environment.

National

Fire behavior models, specifically rate of spread, need to be updated to reflect changes in fuel conditions and observed fire behavior. For example, plume-dominated fire behavior is extremely difficult to predict and model as seen throughout the Station Fire. As a result, the Agency will explore new and emerging technologies that can aid in modeling plume-dominated fires.

The Agency is reinforcing the intent of the Fire Spread Probability (FSPro) model and its current limitations as part of the Wildland Fire Decision Support System (WFDSS) training package.

- The Fire Spread Probability model is currently being enhanced with the addition of gridded forecast weather for short term fire forecasting (3-5 days) and intermediate term fire forecasting (6-10 days). This is a great leap forward in the accuracy and reliability of fire forecasting and modeling.
- Prior to these enhancements, FSPro used the historical range of past weather occurrences to plot the possible spread of a wildfire expressed as percent chance and that nuance was not well understood.

Local- Angeles National Forest (ANF)

The Angeles Forest made the review of current plume-dominated fire modeling and recognition a part of the 80 hour refresher training for the 2010 fire season. Strong emphasis on recognition of factors, weather and burning conditions leading up to plume domination on fires was discussed. The Forest provided our training modules to Los Angeles County Fire for incorporation in their preseason training.

Working with Urban Interface Communities in Volatile Fuels

The fire environment continues to change and people continue to move closer to wildland areas. As a result, the Agency needs to strengthen outreach efforts to communities and stakeholders in these high risk areas and expand the development of joint plans to mitigate risk.

National

The Forest Service has provided considerable support for the national Firewise efforts and National Fire Safe Councils, and needs to continue to encourage all forests with "communities at risk" to ramp up outreach efforts prior to a fire occurring. This has begun on forests throughout the country with a "Continuous Improvement" initiative that greatly emphasizes stakeholder and community involvement for high risk areas.

Local- Angeles National Forest (ANF)

Today there are 19 Fire Safe Councils working with "at risk" communities adjacent to the Angeles National Forest. The Forest is continuing to work with surrounding communities through these Fire Safe Councils to prioritize fuel mitigation, develop evacuation plans, and educate homeowners on steps they can take to protect themselves from catastrophic fires.

DEFENSIBLE SPACE AROUND STRUCTURES

There continues to be confusion around responsibilities for defensible space on private lands. The Forest Service routinely cooperates with state and local government entities that have this responsibility. We recommend this section be combined with the working with urban interface communities lessons learned as this is the platform for sharing the resources available to deal with this issue.

National

The Agency has to better recognize the distinction of fast-moving, highly volatile fuel types and engage communities and cooperators before ignition in planning, evacuations and defensible space where human life and critical infrastructure are at risk.

Building upon Region 5's experience, consider a set of national standards for defensible space on National Forest lands based upon *Firewise* and *Fire Safe Council* principles or state and local standards, if stricter.

Regional

On December 17, 2009 the Pacific Southwest Regional Forester issued a letter directing that the minimum requirement for defensible space around structures within and adjacent to National Forest lands in California be increased from the Federal minimum of 30' to 100'. Utilizing a 100' minimum makes the Forest Service standard consistent with California State and Los Angeles County standards for defensible space. Each Forest Supervisor has the authority to authorize more than 100' clearance, if warranted by site specific conditions. The Angeles NF is currently working with Los Angeles County Fire Department and CALFire to implement this new direction.

Local- Angeles National Forest (ANF)

In addition to complying with the Regional letter of direction, the Forest is currently developing a Defensible Space Zone map that incorporates the 100 foot minimum clearance for structures.

"The WUI Defense Zone is the area directly adjoining structures and evacuation routes that is converted to a less-flammable state to increase defensible space and firefighter safety. This is a strip of land where planned suppression activities involve both containment of the fire perimeter and protection of structures". *ANF LMP, Appendix K page 81*

This Defense Zone map will include all structures in the National Forest and structures on private land within 200 feet of the National Forest. The Defense Zone will also identify evacuation routes with appropriate defense zone buffers. Calculated acreage can be analyzed for impacts. There will be three focus areas in identifying defense zones:

- Protection of Angeles National Forest and other Federal facilities and infrastructure.
- Protection of private structures by enabling owners to come on to NFS lands for treatments to meet LA County Fire Code.
- Identify and maintain evacuation routes in the forest.

Tracking Aircraft Water Delivery

During the Station Fire, the Forest Service was questioned about the quantity and location of water and retardant drops. This same issue has been raised on other large fires around the country. The Agency should improve methods for tracking and displaying the location of water and retardant drops during wildland fire suppression operations.

National

The Agency will continue to test and evaluate the tracking of airborne water/retardant delivery via GPS to assist in real-time and post-incident evaluation of aircraft use.

The Forest Service is also assessing infrared technology in evaluating water/retardant application; and will explore working with contract pilots to determine feasibility of capturing this data using current capability.

Interagency Agreements

Interagency and multi-jurisdictional fire suppression operations are extremely complex and require clear understanding of agreements regarding operational response. It is a best management practice to ensure these agreements and annual operating plans are appropriately updated to account for any changed conditions and in place prior to fire activity.

National

Cooperative Agreements and Annual Operating Plans (AOP) need to be updated annually. Ensure that Cooperative Agreement and AOPs are reviewed and understood by agency and elected officials.

- Cooperative Agreements and Annual Operating Plans, particularly in California, have been reviewed and updated by the Regional Office and the local Forests ensuring that they are understood by all parties, including elected officials.

Local- Angeles National Forest (ANF)

The Angeles NF Forest Supervisor and LA County have already had discussion regarding modification and improvements of their Cooperative

Agreement and Annual Operating Plan (AOP). The changes will include night flying, training and joint response and use of resources.

Resource Ordering and Status System (ROSS) Enhancements

There were several concerns that arose regarding ordering during the Station Fire – such as duplication and cancelation of key aircraft orders. Federal wildland fire agencies use ROSS to locate available resources and track those resources for initial and extended attack as well as large fire management, in California – other fire response agencies still use other Computer Aided Dispatch (CAD) programs for Initial Attack. This can be challenging when working in an interagency environment. During initial and extended attack the need for quick response often promotes using direct communication to get resources mobilized and then "catching it up" in the system after the fact.

National

Determine and take necessary steps to prevent duplication of orders and promote accuracy. Changes are in the works to improve ROSS to respond to mobilizations related to aircraft requests.

- Addition of Contact tab for Aircraft and Equipment resource items – Approved as a high priority and is currently in progress.
- Addition of a unique autodoc for divert of tactical aircraft
- Priority Resource button should flash if there are pending requests in the queue.
- Allow messaging between dispatch centers in ROSS.

Documentation and Record Keeping on Incidents

The Agency received many internal and external inquires regarding the actions taken in the during the early stages of the Station Fire. The Agency's inability to rapidly respond to these requests during and subsequent to fire suppression operations- highlighted the need for improvements in our documentation processes.

National

The need for thorough and accurate documentation cannot be overstated. The Forest Service encourages record keeping at all levels of incident management and will use the Station Fire incident as an example in future training to emphasize the need for improved documentation when using standardized ICS documents such as the 201 and 214.

Develop a fact finding exercise as a developmental training tool for Incident Management Teams and Fire Managers so they have a better understanding of the importance of documentation of decisions, such as the Key Decision Log, and other communications throughout an incident.

Review Procedures

The Forest Service routinely conducts internal reviews to identify potential improvements in programs and policies. These reviews need to be appropriately structured to encompass the complexity of the incident.

National

While the Review of Initial Attack provided important data and met the stated objectives, in retrospect the report should have looked into additional factors including asset utilization.

Standard Operating Procedure for the Forest Service is to use lessons learned and facilitated learning analysis after catastrophic incidents, but we recognize that we need to improve upon this process. As a result, future Forest Service after-action review panels will include a wider range of voices, including qualified individuals who were not in any way involved in the action.

Also, standardizing fact finding methodology and clearly defining what needs to be reviewed in an expansive incident would provide better understanding of outcomes.

During an incident that has the potential of becoming "nationally significant", immediately assign a Liaison to the local unit to assist with fact finding.

Incident Communications, Public Relations and Freedom of Information Act (FOIA)

Timeliness of the release of documentation of records and communication with the media regarding the outcomes of the Station Fire has been questioned. The following have been identified as potential actions.

National

Work with the public information staff of Department of Agriculture and at every level of the Forest Service to improve responsiveness; ensure transparency; and provide timely, accurate information to the public and the media.

Develop a better tracking and communication/coordination protocol between the Washington Office, Regional Office and Forest FOIA staff.

Standardize process for publicly posting FOIA requests on the web.

Follow FOIA guidelines for release of information for all requests. Shortcutting normal protocols lead to confusion and misunderstandings of what has been released and what hasn't; this includes ensuring all documents have been properly redacted and cleared through the proper officials.

Local- Angeles National Forest (ANF)

The issue of media and public scrutiny has been discussed with LA County Fire Department. As a result of the Station Fire, both organizations have agreed to work together in developing joint information processes to address the coordination of incident information, questions and positions regarding response to fires in proximity to each other's jurisdiction.

Dispatching/Emergency Communication Center Protocol

Standardization of dispatch and communications is important during all fires. During the Station fire, there were questions raised about how dispatchers communicated and what resources where ordered. Established best management practices by geographic area could reduce lack of standardization and questions.

National

Explore standardization of dispatch center radio recordings nationally. This should include how frequencies are recorded, how long they should be archived and which channels are to be recorded.

Standardize operating procedures and after hour protocols for when dispatch or coordination center closes.

Standardize how Geographic Area Coordination Center (GACC) tentative airtanker assignment documents need to be prepared, shared, and retained.

Standardize how or if 'holds' should be placed on orders. This common practice can lead to misunderstandings regarding if the order was even placed or acted upon. (see ANF LL below)

Develop protocols to document interagency decisions at the GACC level when incidents occur outside of Multi-Agency Coordinating (MAC) groups' mode.

Identify common language in radio protocols. This would include eliminating the use of local identifiers while on incident with identified positions- e.g. DIV1 was IC at different times but on radio and in records was called "DIV1" or "Chief 1" or "Chief". This makes it difficult to keep track of who is who during incident.

Date/time stamps for WildCAD, dispatch clocks, and radio frequency recording need to be synced.

Continue to field test ground resource tracking capabilities and archiving data similar to Flight Following.

Explore modernization of current frequency recording equipment and procedures (digital commo...voice over IP, auto transcription, and location of transmission).

Explore need to record radio frequencies that are used by Incident Management Teams (IMTs) during large fire management.

Set standards for use of "off-line" communications, i.e.; crew net frequencies, cell phones, texting, etc. while on an incident.

Develop a national or regional standard that addresses how and what should be debriefed when dispatches transition from one shift to another. (avoids duplication of orders, missed orders and increases SA)

Regional

Share Angeles Lessons Learned with all Emergency Communications Centers in Region.

Institute a Standard Operating Procedure (SOP) that all ECCs and GACCs submit all orders when requesting resources.

Local- Angeles National Forest (ANF)

The Angeles National Forest will continue to use the "Closest Forces" concept to request additional resources as described in the California Mobilization Guide.

The Angeles National Forest currently uses and will continue to utilize mutual aid agreements to obtain available resources such as aircraft, handcrews, equipment and key overhead resources from local fire agencies.

The Angeles Emergency Communications Center will relay and document all delays of resources to incident commanders so they may amend or alter plans of attack. (Implemented 2/1/2010)

The Angeles National Forest will not make any "informal" requests for resources. All requests will be entered into the Resource Order and Status System" (ROSS) and placed either to cooperators with mutual aid agreements that may be able to provide the resource or to the Geographic Area Coordination Center (GACC).

This will provide a documentation record of what requests are declined or filled and by whom. (Implemented 2/1/2010)

The Angeles National Forest will document when and why resources are released from an incident, such as by the request of a providing agency, they are excess resources, etc. This will be documented by the Incident Commanders as well as the Dispatch Center. (Implemented 2/1/2010)

The Angeles National Forest will continue to record all radio traffic that is transmitted to the Emergency Communications Center on Forest Net, Forest Repeat and Admin Net. The Angeles is in the process of upgrading its Dictaphone server to be able to record additional frequencies which will include Aircraft Flight Following and Law Enforcement frequencies.

The Angeles National Forest has been reminded that the ECC emergency phone lines are for emergency business. Use of these phone lines for other than emergency business is not authorized. Phone conversations must remain professional and related to business only.

Transitional Briefings- Transfer of Command

Confusion regarding what resources had been ordered and/or released on the Station Fire a seemed to be commonly questioned. Much of this was the need to improve briefings during transitions or transfer of command. As a result the following actions have been recommended.

Local- Angeles National Forest (ANF)

Incident Commanders (IC) will ensure that an accounting of all resources on the incident, released from the incident, and ordered resources is relayed to the incoming IC regardless of complexity. This information will also be validated through the Emergency Operations Center. This will confirm that a seamless transition has occurred and all resources are accounted for.

All Incident Commanders will follow the guidelines that are spelled out in the 2010 Interagency Standards for Fire and Fire Aviation Operations Guide.

Copies of all the Incident Commander's documentation will be given to the incoming IC. The outgoing IC will also retain a copy of the documentation that was given to the incoming IC.

Incident Commander Type III Refresher

On the Station Fire, questions regarding decisions and actions during initial and extended attack were raised. All firefighters are qualified for the positions they are asked to perform. Lessons learned are routinely incorporated in Agency refresher trainings.

Regional

The Agency review of the initial attack events reinforces the value of Type III Incident Commander Time/Pressure Simulations for Qualification Certification.

Local- Angeles National Forest (ANF)

Annually, the Forest will continue to use Time Pressure Simulations for all Type III Incident Commanders. These simulations were completed by all ANF Type III Incident Commanders during the month of April 2010.

The Forest will continue to train with Chief Officers from our cooperating agencies on the Unified Command Process and Procedures.

Interagency Coordination

As stated in the cooperative agreements section, fire response is always an interagency environment, so continued relationship building and knowledge sharing will help ensure there is no breakdown between agencies. Unfortunately during the Station Fire, many of the questions posed by external parties deflated relationships which are critical for future response. As a result additional effort will need to be made to make sure these relationships are repaired.

Regional

Continue to work with individual national forests, to strengthen relationships and communication with cooperating agencies; which includes extensive pre-season communication and planning with stakeholders and communities.

Local- Angeles National Forest (ANF)

Continue to work with all agencies, elected officials, communities and stakeholders to strengthen relationships and communication.

The Forest will work with cooperators in an effort to develop a Joint Information System (JIS) with area cooperators to ensure there is consistent and timely internal and external communications before and during a fire.

In the future, the Forest will evaluate the need for a daily joint Agency Administrator meeting to make sure leader's intent and the intended fire strategy is being met.

INDEX

E

F

G

H

I

L

M

N

O

P

R

S

T

U

V

W

Z